T0135201

Intelligent Systems Reference Library

Volume 83

About this Series

The aim of this series is to publish a Reference Library, including novel advances and developments in all aspects of Intelligent Systems in an easily accessible and well structured form. The series includes reference works, handbooks, compendia, textbooks, well-structured monographs, dictionaries, and encyclopedias. It contains well integrated knowledge and current information in the field of Intelligent Systems. The series covers the theory, applications, and design methods of Intelligent Systems. Virtually all disciplines such as engineering, computer science, avionics, business, e-commerce, environment, healthcare, physics and life science are included.

More information about this series at http://www.springer.com/series/8578

Constantinos Artikis · Panagiotis Artikis

Probability Distributions in Risk Management Operations

Constantinos Artikis
Department of Informatics
University of Piraeus
Piraeus
Greece

Panagiotis Artikis
Anglia Ruskin University
Cambridge
UK

ISSN 1868-4394 ISSN 1868-4408 (electronic)
Intelligent Systems Reference Library
ISBN 978-3-319-36202-1 ISBN 978-3-319-14256-2 (eBook)
DOI 10.1007/978-3-319-14256-2

Springer Cham Heidelberg New York Dordrecht London

Softcover reprint of the hardcover 1st edition 2015

Printed on acid-free paper

Springer International Publishing AG Switzerland is part of Springer Science+Business Media (www.springer.com)

Foreword

Many terms have been used to describe our society. One of them, which appeared in the 1980s and has become popular since the 1990s, is the term *risk society*. By risk society we usually understand a society which is preoccupied with the future and, thus, has to deal with the risks associated with it.[1]

There are many sources of risk for our society and mankind: environmental change or disaster risks, risks from large-scale accidents, risks from pandemic disease and biological war, risks from large meteorites, risks from civil unrest, even risks from extraterrestrial or artificial intelligence. Although some sociologists are pessimistic about the ability of our society to address such risks, Giddens is more optimistic, suggesting that there "can be no question of merely taking a negative attitude towards risk. Risk needs to be disciplined, but active risk taking is a core element of a dynamic economy and an innovative society."[2]

For some time now, the community of academic and applied risk managers has been showing a very strong interest in stochastic models. This stems from the fact that the availability of stochastic models of real-world problems allows the intelligent analysis of risk and planning for its management. As the environments are very complex, in which risk managers are required to work, the corresponding stochastic models are equally complex.

In the book at hand, the authors investigate such stochastic models making use of the results of the theory of mixed probability distributions and characteristic functions. The book is addressed to all graduate students and researchers as well as to practitioners of risk management. As such, it is self-contained, requiring only an introductory knowledge of probability and stochastic models.

---

[1] Giddens, A. (1999). Risk and responsibility. *Modern Law Review* 62(1), 1–10.

[2] Giddens, A. (1999). *Runaway world: How globalization is reshaping our lives*. Profile, London p. 29.

I believe that the authors have done a good job at addressing the tackled issues. I consider the book a good addition to the areas of stochastic modelling and risk management and I am confident that it will help graduate students, researchers and practitioners to understand and expand the use of stochastic models in real-world problems.

George A. Tsihrintzis
University of Piraeus, Greece

Preface

The severe unstable performance and evolution of large organizations of the risk society is generally recognized by management experts and the corresponding academic community. This recognition significantly supports the amplification of the contribution of risk management in the performance and evolution of modern complex organizations. As a consequence, the discipline of risk management is in the ascendancy. Moreover, risk management has a long and rich history matching the breadth that characterizes this discipline. During the last 20 years, the interest of the academic community in the discipline of risk management has been considered very important. The presence of that interest follows from the plethora of conferences and research papers related to risk management matters, implemented by experts on investigating and applying concepts, operations and principles of that discipline. The recognition of risk management as a powerful tool for solving problems arising in a very wide variety of natural and human activities is general. Within the risk managers and the academic risk management community, there has been a very strong interest in stochastic models. Once stochastic models are available for the description of real-world problems in the area of risk management, it is possible to intelligently evaluate the issues and alternatives and chart courses of action for a proactive risk management program, which is particularly important for implementing the strategic goals of an organization. Such models are employed in many areas of the risk management process because risk managers work in an extremely complicated and uncertain environment. From the fact that stochastic models give their users a chance to isolate and study the various thought processes involved, risk managers can gain insight into how to improve their decision-making process in developing determinations about the risks faced by an organization. Risk identification, risk measurement and risk treatment constitute the fundamental risk management operations of an organization. The effective application of planning, organizing, staffing, directing and controlling in risk management operations implies the effective performance and evolution of modern complex organizations. The stochasticity of the risk management process is an inevitable result of the presence of random factors in the fundamental quantitative components of risk and the fundamental risk management operations of an organization. The handling of these random factors requires the risk managers of modern large organizations to have strong capabilities in formulating and implementing complex stochastic models for describing fundamental concepts and operations of risk management. The consideration of stochasticity of the risk management process is an extremely important structural element for developing proactive risk management programmes. The recognition of probability theory as a powerful analytical tool of risk management constitutes a very important reason for undertaking research activities in the area of probability distributions arising in stochastic modelling of concepts and operations of risk management. The investigation of such stochastic models makes use of the results of the theory of mixed probability distributions. In particular, the very strong results of the theory of characteristic functions corresponding to mixed probability distributions are extremely useful for investigating properties and applicability in risk management operations of stochastic models of this kind.

The authors would like to thank Profs.-Drs. Maria Virvou and George A. Tsihrintzis of the University of Piraeus, Piraeus, Greece, for their encouragement, fruitful discussions and comments over the last few years, which have helped to shape this monograph. They would also like to thank Prof.-Dr. Lakhmi C. Jain of the University of Canberra, Canberra, Australia, and the University of South Australia, Adelaide, Australia, for agreeing to include this monograph in Springer's *Intelligent Systems Reference Library* series that he edits. Finally, they would like to thank Springer and its personnel for their wonderful work in producing this volume.

Constantinos Artikis
Panagiotis Artikis

Contents

Chapter 1
Fundamental Concepts of Risk Management

Abstract Risk management may be defined as the systematic process of managing the risks threatening an organization in order to accomplish its goals in a way consistent with common interest, human protection, environmental factors and the law. It consists of the planning, organizing, directing, the safety operations or equivalently the risk management operations with the aim of developing an efficient plan that decreases the negative results of risks threatening that organization. The first chapter consists of three parts. The first part concentrates on the definition, historical consideration, components, consequences, and cost of risk. Moreover, the second part concentrates on the operations, goals, structural disciplines, essence, ascendancy, systemic approach, cindynic consideration, philosophy, and the fundamental factors of the evolution of risk management. The third part presents the theoretical and practical possibilities of stochastic modeling and stochastic models for shaping risk management as a particular discipline of general management.

1.1 Introduction

Recent social theory points towards the evolution of post-industrial society, termed risk society. This phenomenon is impacting on the operating environment of many organizations forcing on them dynamic and markedly unpredictable change. Such instability has been recognized by many managers and risk management is becoming an increasingly common term in business life. The unstable evolution of the organizations of risk society is recognized by the management experts and the corresponding academic community. Such recognition substantially supports the amplification of the role of risk management in the evolution of modern large organizations. As a result, the discipline of risk management is in the ascendancy. Moreover, risk management has a long and rich history matching the breath that characterizes this discipline. During the last two decades, the interest of academic community for the discipline of risk management is generally recognized as particularly significant. The existence of that interest follows from the plethora of

C. Artikis and P. Artikis, *Probability Distributions in Risk Management Operations*,
Intelligent Systems Reference Library 83, DOI 10.1007/978-3-319-14256-2_1

conferences and scientific publications, on risk management problems, implemented by experts on the investigation and application of concepts, operations, and principles of this discipline. The admission that risk management constitutes a very strong tool for solving problems, arising in a wide variety of natural and human activities, is universal. In particular, risk management activities are as significant for the contemporary service economy as the textile industry was at the beginning of the industrial revolution. Risk management cannot be implemented by the risk manager only or some other manager of an organization. Such an implementation requires the very good collaboration of all the personnel of an organization. Risks arising from the production activities of an organization require the collaboration of the production managers for the guidance of the involved personnel in the development of safe goods and services with safe procedures. Similarly, risks arising from the promotion of goods and services require the collaboration of marketing managers for the guidance of the involved personnel in order to avoid actions which can be considered as conditions or causes of various risks. Thus, the function of the executive charged with responsibility for risk management is not to personally reduce the adverse effects of all risks but, instead, to coordinate the efforts of all managers in reducing risks for which each of them has some responsibility and, therefore, control. That means the existence of a substantial difference between the role of the risk manager and the roles of the other managers of an organization. The basic characteristic of the role of risk manager is the coordination of all managers of an organization for the effective treatment of risks threatening the existence, function, and evolution of that organization. These characteristics are substantially amplified by treatment of risks threatening the existence of an organization.

The relationships between the risk manager and other managers at all levels of the organization are reflected in the communications flowing to the risk manager from other departments. Three principal types of communications should reach the risk manager. First, the risk manager should receive regular reports on changes in the risks threatening the organization, or on situations from which changes can be reduced. An important staff function of the risk manager might well be to design a procedure by which all relevant reports reach the risk management department. Second, the risk manager should receive guidance from senior management on risk management policy, preferably as a written statement covering at least the broad goals of the risk management process, the relationship of this process to the other processes within the organization, and the standards for acceptable performance of the risk management department. It is quite obvious that the guidance of senior management to the risk manager very clearly defines the risk management policy of an organization. In practice, the risk manager often drafts or participates actively in drafting this statement of risk management policy to be submitted to and approved and issued by senior management. Third, perhaps as part of the risk management policy of the organization, the risk manager should receive a grant of staff and perhaps limited line authority. The third type of information having as receiver the risk manager concerns the delegation of authority to that manager. The delegation of this authority is realized for supporting the development and implementation of the risk management policy of the organization. The three types of information that the risk manager receives from the communications with the departments of the organization make absolutely explicit that the risk manager can contribute to the evolution of an organization with the properties of receiver, user, creator, administrator, transmitter, and organizer of information. These properties are extremely important for developing and implementing of a very effective risk management policy.

Risk management is mainly a logical process of making decisions. The practical situations incorporating risk management problems are innumerable. The consideration of the risk management process as a logical process of making decisions makes the risk management process a completely exceptional process. The possibility of applying a logical process of making decisions in every situation incorporating risks has as a direct consequence the separation of the role of the risk manager from the role of any other manager of an organization. It is generally recognized that the contribution of risk manager in implementing the goals of an organization is particularly important. This is mainly due to the exceptional importance fact that the risk manager deeply knows the function of every department of an organization and thus the risk manager is able to have a holistic and systemic perception of that organization. This perception increases the credibility of the role of the risk manager in defining the strategic goals and the risk management policy of an organization. This credibility becomes of extremely importance if the risk manager has the ability and the experience of formulating mathematical models for describing and solving risk management problems. The risk management process or equivalently the logical process of decision making in every situation incorporating risks has similarities with the process, which is known as scientific method. The application of this method from a scientist means the definition of a real problem, consideration of alternative assumptions which can terminate in the solution of this problem, selection of the most probable assumption, verification of this assumption, and comparison of the results derived from the solution of that problem with the real results, which could arise, if the selection of the assumption was correct. If the results of the solution of the problem are identical with the real results then the selection of the assumption is correct and thus the problem has been solved. If there is disagreement between the two categories of results then a new assumption must be verified. The application of a logical process for describing and solving problems in the area of risk management has the three following significant advantages. First, since the operations of the risk management process have many similarities with the techniques used by the academic community and the management experts for solving problems, then a risk management expert has the possibility to apply these techniques for solving problems in the area of risk management. Second, the application of the logical process in approaching new problems gives to risk management experts the possibility to direct themselves to the best use of their available time and the effective use of the resources of an organization. The third advantage of a logical process for making decisions is the very significant support that such a process provides in explaining and justifying the risk management decisions to the senior and other managers. A logical process of making risk management decisions provides an explicit frame of presenting problems and solutions of risk management problems with such a way that every member of the organization understands, respects, and supports. The consideration of risk management process as a logical decision making process in situations incorporating risks makes quite clear the necessity of formulating and applying mathematical models in various risk management operations. The presence of random factors in these operations supports the formulation of stochastic models as strong analytical tools in the discipline of risk management. The present work mainly concentrates on the formulation, investigation, and implementation of stochastic models for risk measurement and risk treatment operations. In particular, the present work is devoted to the establishment of conditions for explicitly evaluating the probability distributions of such models.

Within the international scientific risk management community, there has been a very strong interest in stochastic models. Such models are employed in many areas of risk management because risk managers work in an extremely complicated and uncertain environment. From the fact that stochastic models give their users a chance to isolate and study the various thought processes involved, risk managers can gain insight into to improve their decision making process in making determinations about the risks faced by an organization. Quantifiable information about the risks to which an organization may be exposed is a necessity if risk managers are to make prudent decisions for proactive and reactive treatment of these risks. Stochastic models are extremely important in analyzing such information. The mystique which has grown up around the application of stochastic models in risk management problems has obscured for many risk management experts the extent to which they can really apply such models in their daily work. The more sophisticated stochastic models are very efficient elaborations of what most risk management experts instinctively do in reaching and defending their more practical management decisions. Management skill is necessary for efficient treatment of risks faced by an organization. However, without a thorough analysis using the best stochastic models currently available, risk management becomes virtually impossible. Risk managers believe that the processes of making and implementing effective decisions for a wide variety of risk management problems must be based on stochastic models arising as results of hard and long standing activities. It is generally accepted that the management abilities of risk experts are necessary for the successful implementation of risk management principles. However, it is also generally accepted that the implementation of risk management principles cannot be successful without the formulation and implementation of effective stochastic models. A formal treatment of the risk management process relies on the more general problem of decision making under conditions of uncertainty, which has traditionally approached with the theory of expected utility. The formal theory of expected utility is, however, too restricted to handle practical applications, but it can substantially contribute to give very significant insights by directing attention to the validity of underlying assumptions. Decision making under conditions of uncertainty heavily relies on stochastic models for predicting outcomes of given actions and stochastic models are often used to describe the problem of decision making under conditions of uncertainty. The decision making process, under conditions of uncertainty, uses stochastic models in order to make predictions about the outcomes of given actions of the risk experts of an organization. Risk managers and experts in formulating and implementing stochastic models are frequently called upon to predict the occurrence of uncertain events. The predictions of such events usually form the basis of extremely important risk management decisions. This means that an understanding of uncertainty is critical to the performance of the risk management process. Moreover, it is important that risk managers accurately reflect and communicate uncertainties. For many years the risk management community has recognized probability theory as one of the most important representations of uncertainty. This implies that stochastic modeling and stochastic models have their proper place in the area of risk management practices. With stochastic models, the risk manager can obtain valuable information concerning selection and implementation of risk management operations.

Once such models are available, it is possible to intelligently evaluate the issues and alternatives and chart courses of action for a proactive risk management program. Many stochastic models have been readily available to corporate risk managers for making risk management decisions. These models are being used to improve not only the decisions are made, but also in the presentations to top management of critical issues. As risk managers begin to use stochastic models more skillfully they will become a more significant part of the long range planning team.

1.2 Historical Consideration of Risk

The diachronic approach of the concept of risk from man includes the following periods.

Period of Blood

Period of Tears

Period of Neurons

The study of this approach constitutes a significant factor for developing, evaluating, selecting, and implementing new methods in the disciplines of risk management, crisis management, and cindynics.

1.3 Period of Blood

During this period, the therapy for natural and man made risks had to be found in animal or human sacrifices. The act of pouring blood of animals or pure young girls or boys was considered as more efficient than what we call today meetings of risk managers with brokers and insurers. More precisely, this act was thought to reduce the probability of return of such risks.

1.4 Period of Tears

With the spread of Christianity, human sacrifices will give way to other ritual practices. This will start the period of tears where spilling tears instead of pouring blood with the invocation of divine providence. Prayers and processions will be regarded as natural treatments against risks. These ritual practices for risk will perpetuate the idea that God and not man pulling the strings of risk.

1.5 Period of Neurons

This period will start from a dispute between Voltaire and Rousseau, on the occasion of the earthquake of Lisbon in 1755. Since the time of Rousseau begins the period of neurons, during of which the very nature of the issue of risk will change. The risk will emerge from the mythological space to enter the logical space. The period of neurons incorporates the following phases.

Philosophical Phase

Technological Phase

Scientific Phase

1.6 Philosophical Phase

Having shed blood and tears, man will undertake responsibility and put his mind in front of risk. Indeed, while Voltaire continues to accuse nature and divine province, Rousseau observes that the decisions of building cities in seismic zones implicate intelligence and responsibility of man. The observation of Rousseau is the beginning of philosophical discussion concerning risk. That discussion continued until the start of the Second World War.

1.7 Technological Phase

This phase begins after the start of the Second World War and continued until the last decade of the previous century. Characteristics of this phase are the determination of quantitative rules on the frequencies of aviation accidents, studies of reliability of electronic equipment, and the emergence of risk management.

1.8 Scientific Phase

This phase begins in the last decade of the previous century and continuous today. During this phase, the risk is the subject of science. The study of risks will result in the gradual accumulation of facts and concepts. In the last decade of the previous century cindynics, the science of risk, coming from a fetal situation will come to light and will grow very quickly.

Meanwhile, man will get a more reactive position by rejecting the myth of destiny accidents and catastrophes. He will admit that he is able to greatly reduce the effects of earthquakes, deaths in cities, deaths of transport, and ecological destruction. Quite simply, man will feel responsible by rejecting the misleading utopia of zero risk and engage in strategies of the calculated risk. From these calculations, risk treatment techniques will witness a development without precedent.

From the moral of destiny man moved to the moral of responsibility. In this wavering, man is ready to recognize into the accidents the confirmation of certain deficits of his knowledge, precautions or his actions. We are in a convergence between a moral leap of man and an opening of scientific approaches to risk.

A simultaneous acceleration of research and publications on cindynics and ethics characterizes the decade 1980–1990. This synchronization is based on a number of major accidents of this decade.

1.9 Economic Loss and Economic Benefit

In this work, the concept of loss is generally identical with the concept of economic loss, or equivalently the loss which can be interpreted in economic terms. Many losses cannot be easily interpreted in economic terms and a very significant problem of risk management is the invention of procedures for the economic interpretation of various losses. Such procedures frequently incorporate subjective factors. Correspondingly, in this work, the concept of benefit is generally identical with the concept of economic benefit.

1.10 Pure Risk

In risk management the concept of pure risk is related to the concept of random economic loss, or equivalently the economic loss whose, the size and the time of occurrence cannot be predicted with certainty. The following sections make use of the term “risk” instead of the term “pure risk”.

The concept of risk has particularly important applications in many theoretical and practical disciplines. This means that a general definition of risk is not possible. Modern risk management considers the concept of risk as an initial one. The relevant bibliography recognizes that the concept of risk is very complicated and the understanding of this concept requires the use of a set of other qualitative and quantitative concepts. These concepts are called components of risk. The set of the components of risk is not precisely defined. However, it is generally accepted that the fundamental elements of the set of the components of risk are the following concepts.

Risk Cause

Risk Severity

Risk Frequency

Risk Duration

Hazard

Many practical investigations have reached to the conclusion that these five concepts are particularly useful in treating of a wide and significant variety of problems incorporating the concept of risk.

1.11 Risk Cause

A random event which can give rise to a loss of random size in random time is defined as risk cause. The appearance of a risk cause is called a risk occurrence. Examples of risk causes are the following.

Fire

Storm

Explosion

Theft

Vandalism

Negligence

Strike

War

Earthquake

1.12 Risk Severity

The size of the economic loss arising from the occurrence of a risk is defined as the severity of that risk. The suitable stochastic model for describing and investigating the severity of a risk is a positive random variable.

1.13 Risk Frequency

The number of occurrences of a risk in a given time interval is defined as the frequency of that risk. The suitable stochastic model for describing and investigating the frequency of a risk is a random variable taking values in the set of nonnegative integers.

1.14 Risk Duration

The length of the time interval, into which the cause of a risk is active, is called duration of that risk. The suitable stochastic model for describing and investigating the duration of a risk is a positive random variable.

1.15 Hazard

Any condition contributing to the appearance of the cause of a risk is defined as hazard. There are three types of hazards.

Physical

Moral

Morale

1.16 Physical Hazard

Physical hazards arise from the activities of nature. Examples of physical hazards for the risk cause "collision of car" are the following.

Icy Road

Slippery Road

Insecure Car

Restricted Visibility

1.17 Moral Hazard

Moral hazards arise from dishonesty or character defects in an individual. Examples of moral hazards for the risk cause "collision of car" are the following.

Unstable Personality of Driver

Inadequate Training of Driver

Arrogant Behavior of Driver

1.18 Morale Hazard

Morale hazards arise from carelessness or indifference to a risk. Morale hazard should not be confused with moral hazard. Morale hazard refers to individuals who are simply careless about protecting their property. Moral hazard is more serious since it involves unethical or immoral behavior. Examples of morale hazards for the risk cause "collision of car" are following.

Young Age of Driver

Damaging Tendency of a Driver

1.19 Fundamental Stochastic Components of Risk

The set of the fundamental stochastic components of risk includes the following elements.

Risk Severity

Risk Frequency

Risk Duration

The role of the fundamental stochastic components of risk in formulating stochastic models of the risk management discipline is generally recognized as extremely important.

1.20 Consequences of Risk

The consequences from the occurrence of a risk are the following.

Economic

Social

Political

Psychological

Natural

Legal

The present work concentrates on the economic consequences of risk. Since the separation of the economic consequences from the other consequences of risk is not possible then it is absolutely necessary the consideration of the economic consequences of risk to incorporate some attention to the other consequences.

1.21 Reaction to Risk

The reaction of a person to risk is the way in which the person behaves in a situation incorporating risk. One factor affecting this reaction is the uncertainty of the person. Other things being equal, one would expect the person to react more strongly, either positively or negatively, the greater the uncertainty of this person. Other factors that may be equal or greater significance are the size of the potential benefits or losses involved and the impact of these benefits or losses upon the economic status of the person. Even if all these factors are the same, people react differently because their personalities, as determined by their heredity and their environment, vary. Indeed, the same person may have a different affinity for or aversion to risk at different ages and in different situations. Individuals making decisions under conditions of risk should be aware of the impact of their own risk attitudes upon their decisions. Upon closer inspection they may decide to change their attitudes. If they are making decisions on behalf of a family or an organization, they should examine the extent to which they should accept the attitude of others. Persons delegating these decisions to someone else also study the attitude of this person toward risk and how it affects the decisions that this person makes. In many cases it may be suitable to specify the attitude that should be assumed in making decisions. Many researchers have investigated the demographic characteristics, personality traits, and environmental conditions that may determine the reaction of a person to risk. These investigations have substantially contributed to the understanding of how persons behave in situations incorporating risk. They suggest that such behavior is extremely complex, depending upon a host of factors and varying over time. They also indicate that a person may react differently to financial risks than to physical and social risks. Investigations that have attempted to describe the reaction of a person to risk in terms of one demographic or personality trait have generally yielded contradictory results. For example, some studies suggest that women tend to be more averse to risk than men. Other studies suggest no difference in risk aversion between men and women. Similar contradictory evidence exists concerning the effects of intelligence, and of education. One extremely interesting and important finding is that individuals tend to be more willing to accept risk after they participate in a group facing the same risk than they would have previously as individuals. Consequently group decisions tend to be riskier than the average decision made by the members of the group prior to their group experience. The most popular explanation of this risky shift is that individuals view themselves as being at least as willing as their peers to undertake risks. Recent experiments, however, suggest that groups do not always respond in this way. Indeed, under certain conditions group decisions may be more cautious than the average individual decisions.

1.22 Flight-Fight-Freezing Syndrome

The retreat in front of risks beyond our forces is a logical act with psycho-emotional consequences. The trend of flight from adverse situations or unwanted persons was and continues to be a primordial, life-saving knowledge and human ability of the seasons that man was still naked and weak having to face animals more bigger than him in the daily struggle for survival on our planet. The classic syndrome was and remains almost unchanged and identified with the dilemma of "flight or fight". The fundamental psychological question which differentiates the healthy from the pathological trend of flight is associated with both the severity and the frequency of risk with which a particular person chooses to flee from adverse situations or unwanted persons.

The modern conditions of life, without imitating the brutal challenge of the jungle, contain significant elements of risk, creating numerous symbolic conflicts between individuals and groups, looting continually the personality and mental balance of man. Of course modern man, as a derivative of the industrial society, has the necessary defense mechanisms of function and survival under the requirements of the psycho-socio-economic structures within which he lives and works.

The trend of flight, innate part of the classic syndrome of "flight-fight" which is present in every person, is defined as problematic when it becomes the only way to tackle the demands of everyday life of a person. This finding is also valid for the trend of fight.

The modern literature of risk management proposes the trend of freezing, as the third innate part, in the classic syndrome of "flight-fight". The trend of freezing means the acceptance of a waiting period before the manifestation of the trend of flight or the trend of fight. Therefore, the syndrome of "flight-fight-freezing" is the modern expression of the classic syndrome of "flight-fight".

1.23 Structural Elements of Firm

The structure of a firm includes the following elements.

Human Resources

Emotions, desires, wishes, initiatives, dynamic and creative thinking, values, habits, and the motives of human resources of a firm constitute the regulating factors of its behavior.

Production Facilities

This structural element arises as composition of buildings, machinery, and production processes of the firm.

Raw Materials

The purpose of raw materials is the supplying of the production facilities of the firm.

Energy

This element constitutes the operative force of the firm.

Money

This element constitutes the medium of exchange for the supplying of production facilities, cost of labor, and purchase of raw materials and energy.

Information

This element refers to the factors associated with the internal and external world of a firm.

A detailed consideration of the elements shaping the structure of a firm constitutes a very significant factor of the successful application of risk management principles.

1.24 Objectives of Firm

The objectives which establish and justify the existence of a firm are the following.

Production of Goods or Services to Meet Needs

Meeting the Purposes of the Members of the Firm

The production of goods or services from the firm is the means to meet the aspirations of all members of the firm, or equivalently of the individuals directly associated with the firm and wishing the meeting of their needs.

The existence of the firm constitutes the main motive of the behavior of the firm. The survival of the firm is not only ensured by the production of goods or services, but by the ability of the firm to meet the needs of its members.

The contribution of risk management in the implementation of the objectives of a firm is extremely important.

1.25 Classifying Operations of Firm

A classification of the operations of a firm with particular practical importance is the following.

Technical Operations

The exclusive purpose of these operations is the production of the goods and services of the firm.

Commercial Operations

These operations are designed for marketing of goods and services produced by the firm.

Financial Operations

These operations are devoted to the description of the role of money in the production process of the firm.

Safety Operations

The main purpose of these operations is the reduction of damages due to the activities of the firm.

Accounting Functions

These operations are designed to systematically record all the financial transactions of the firm.

Management Operations

These operations seek the most efficient combination of the resources of the firm to achieve its objectives. Management operations are also called operations of management activity. The basic characteristic of management operations is the application of these operations to the technical, commercial, financial, safety, and accounting operations of the firm.

1.26 Safety Operations

The set of safety operations is not precisely defined. However, it is generally recognized that fundamental elements of this set are the following operations.

Risk Identification

Risk Measurement

Risk Treatment

It is well known that the term "risk management operations" is frequently used instead of the term "safety operations". The present work makes use of the term "Risk Management Operations".

1.27 Risk Identification

Risk identification is the operation by which an organization systematically and continuously identifies property, liability, and personnel exposures as soon as or before they emerge. The implementation of risk identification is based on the following two activities.

Development of a list incorporating all the risks that could occur to any organization;

Application of that list for identifying those risks threatening the given organization.

The risk manager may personally conduct this two steps procedure or may rely upon the services of an insurance agent, broker, or consultant. Risk identification is the first and most important safety or risk management operation. Moreover, practical experience supports that risk identification is the most basic and time consuming activity of the risk manager. The following factors have particular role in the successful implementation of risk identification.

Statistical Data

Imagination

1.28 Risk Measurement

Risk measurement is recognized as the most obscure and complex risk management function. The obscurity and complexity of this operation is due to the inability of providing a definition of the concept of risk which definition is accepted by all disciplines incorporating that concept. From a strict mathematical point of view, risk measurement is directly related with the concept of total risk severity. The total risk severity is defined as the sum of losses from the occurrences of a risk into a given time interval. Thus, the concept of total risk severity has as structural elements the concept of risk severity and the concept of risk frequency. Since risk severity is modeled by a positive random variable and risk frequency is modeled by a discrete random variable with values in the set of nonnegative integers then total risk severity is frequently modeled by a random sum of positive random variables. Such a stochastic modeling of total risk severity is based on the absence of time value of money. If time value of money under conditions of uncertainty is considered then the total risk severity is modeled by a random sum, of positive random variables, which has a very complex form. Consequently, the formulation of a random sum for the present value of the total risk severity under conditions of uncertainty for risk severity, risk frequency, time value of money, and risk occurrence time is defined as the strict mathematical consideration of risk measurement. The evaluation and the investigation of the probability distribution of such a random sum constitute the most important part of the mathematical consideration of risk measurement. The extreme difficulties of formulating this random sum and the evaluation of the corresponding probability distribution very frequently result to the avoidance of the strict mathematical consideration of risk measurement. Such an avoidance frequently results to wrong interpretation of the risk measurement operation. The advantages of stochastic modeling substantially support the strict mathematical consideration of risk measurement. The formulation and investigation of stochastic models, supporting the strict mathematical consideration of risk measurement, constitute two of the main purposes of the present work.

1.29 Risk Treatment

The risk manager, after the identification and measurement of the risks threatening an organization, must apply the operation for treating of these risks. Risk treatment is a complex operation incorporating the following operations.

Risk Control

Risk Financing

Risk treatment mainly includes decisions for handling the risks threatening an organization in a proactive and reactive way. This operation is considered as the most dynamic risk management operation of an organization.

1.30 Risk Control

The risk manager of an organization uses this operation in order to reduce and make more predictable the losses due to the occurrences of risks faced by an organization. Risk control is a complex operation incorporating the following operations.

Risk Avoidance

Control of Total Risk Severity

Separation of Exposures to Loss

Combination of Exposures to Loss

Risk Transfer

The proactive character of risk control is the main advantage of this risk treatment operation.

1.31 Risk Avoidance

One way to control a particular risk is to avoid the property, person, or activity with which the exposure is associated. Risk avoidance can be implemented by the following two ways.

Refusing an Exposure

Abandoning an Exposure

According to the first way the firm refuses to undertake an exposure to risk and according the second way the firm abandons an exposure to risk.

Risk avoidance is considered as the most effective risk control operation. If risk avoidance is implemented by avoiding the factor which is exposed to risk then this operation incorporates the opportunity cost that is the profit which did not occur by losing an opportunity for profit. However, if risk avoidance is implemented by abandoning of such a factor then risk avoidance incorporates the cost of undertaking that factor.

Risk avoidance, whether it is implemented by abandonment or by refusal to accept the risk, should also be distinguished from operations of controlling total risk severity described by the next section. These operations assume that the firm will retain the property, person, or activity creating the risk but the firm will conduct its operations in the most possible safe manner.

Risk avoidance is a useful, fairly common approach to the handling of risk. By avoiding a risk exposure the firm knows that it will not experience the potential losses or uncertainties that the exposure may generate. On the other hand, it also loses the benefits that may have been derived from that exposure. Risk avoidance usually includes the following factors.

Possibility of Applying Risk Avoidance

First, risk avoidance may be impossible. The more broadly the risk defined, the more likely this is to be so. For example, the only way to avoid all liability exposures is to cease to exist.

Losing Benefits of Applying Risk Avoidance

Second, the potential benefits to be gained from employing certain persons, owing a piece of property, or engaging in some activity may so far outweigh the potential losses and uncertainties involved that the risk manager will give little consideration to avoiding the exposure. For example, most businesses would find it almost impossible to operate without owing or renting a fleet of cars. Consequently they consider risk avoidance to be an impractical approach.

Adverse Consequences of Applying Risk Avoidance

Third, the more narrowly the avoided risk is defined, the more likely it becomes that avoiding that risk will increase another risk. For example, a firm may avoid the risks associated with air shipments by substituting train and tracks shipments. In the process, however, it has created some new risks.

To implement a risk avoidance decision the risk manager must define all those properties, persons, or activities that create the exposure the firm wishes to avoid. With the support of top management the risk manager should recommend certain policies and procedures to be followed by other departments and employees For example, if the objective is to avoid the risks associated with air shipments, all departments might be instructed to use train and truck shipments only.

Risk avoidance is successful only if there are no losses from the exposure the firm wishes to avoid. Indeed, the method will not have been properly implemented if some prohibited activity took place but the firm was lucky and no loss occurred. To illustrate, if some air shipments are made in violation of the policy stated above, some correction is necessary even if the firm incurred no losses on those particular shipments. In addition the risk manager should of course reevaluate periodically the decision to use risk avoidance techniques.

1.32 Control of Total Risk Severity

This operation concentrates on the reduction of the total loss from the occurrences of a risk into a given time interval. The control of total severity of a risk is devoted to the reduction of the severity of any occurrence of the risk and also the reduction of the frequency of the risk into the given time interval.

Examples of risk frequency reduction operations are the following.

Tightening Quality Control Limits

Safety Meetings

Improving Product Design

Examples of risk severity reduction operations are the following.

Providing Automatic Alarm Systems

Immediate First Aid

Medical Care

It is readily understood that the successful implementation of the control of total risk severity is substantially facilitated by the simultaneous applications of risk frequency and risk severity reduction operations.

1.33 Separation of Risk Carriers

We consider a system of entities facing the same risk. Generally, an entity facing a risk is called a risk carrier. If some of entities can be detached from the system without impeding the function of the system then we implement the risk control operation which is called separation of risk carriers. The implementation of this operation is based on the effective handling of the concept of time and the concept of space. This means that detaching some entities from the system may have short or long duration and require small or large space.

Examples of applications of separation of risk carriers are the following.

Individual Training of Personnel

Individual Evaluation of Personnel

Dispersion of Parking Places

Assigning of Responsibilities

Selection of Private Transportation

Rejection of Mass Entertainment Events

The separation of risk carriers as a risk control operation is particularly important if the risk carriers are people.

It is quite obvious that the separation of risk carriers aims to reduce the number of the risk carriers which will suffer the adverse effects of the risk faced by these carriers.

The applicability of the separation of risk carriers depends on the kind of risk taken.

1.34 Combination of Risk Carriers

We consider various entities facing the same risk and which do not constitute a system. If some or all the entities can be combined in order to form a system then we implement the risk control operation which is called combination of risk carriers. The combination of risk carriers runs counter to separation of risk carriers.

Examples of applications of separation of risk carriers are the following.

Group Training of Personnel

Group Evaluation of Personnel

Remove of Responsibilities

Unifying Activities of Firm

Group Action

Collective Decision Making

Rejection of Selfish Behavior

The combination of risk carriers, as the separation of risk carriers, is of particular practical importance.

It is quite obvious that the combination of risk carriers aims at enhancing the effectiveness of risk carriers in the treatment of risk threatening these risk carriers.

The applicability of combination of risk carriers, as the applicability of separation of risk carriers, depends on the kind of risk taken.

1.35 Risk Transfer

Risk transfer, as a risk control operation, refers to the various methods other than insurance by which a risk and its potential financial consequences can be transferred to another party. Some noninsurance techniques that are commonly used in risk management programs include contracts, leases, and hold-harmless agreements.

Risk transfer must not be confused with risk insurance which is a risk financing operation.

In a risk management program, noninsurance transfers have several advantages. They are summarized as follows.

The risk manager can transfer some potential losses that are not commercially insurable

Noninsurance transfers often cost less than insurance

The potential loss may be shifted to someone who is in a better position to exercise loss control

However, noninsurance transfers have several disadvantages. They are summarized as follows.

The transfer of potential loss may fail because the contract language is ambiguous. Also, there may be no court precedents for the interpretation of a contract that is tailor-made to fit the situation

If the party to whom the potential loss is transferred is unable to pay the loss, the firm is still responsible for the claim

Noninsurance transfers may not always reduce insurance costs, since an insurer may not give credit for the transfers

1.36 Risk Financing

This operation is complex and includes the following two operations.

Risk Retention

Risk Insurance

Risk financing as a risk control operation is devoted to obtaining the funds for covering the losses arising from the risks threatening an organization. The purpose of risk financing is reactive.

1.37 Risk Retention

The most common operation of risk treating is retention by the organization itself. The source of funds is the organization itself, including borrowed funds that the organization must repay. Risk retention may be active or passive, conscious or unconscious, planned or unplanned. Risk retention, however, is primarily appropriate for high frequency, low severity risks where potential losses are relatively small.

1.38 Risk Insurance

Insurance is one of the basic risk treatment operations, but it is easily the most important illustration of the transfer technique and the keystone of most risk management programs. Insurance can be defined from the two points of view. First, insurance is the protection against financial loss by an insurer. Second, insurance is a device by which the risks of two or more persons or firms are combined through actual or promised contributions to a fund out of which claimants are paid. The distinctive feature of insurance as a transfer device is that it involves some pooling of risks, or equivalently the insurer combines the risks of many insured. Through this combination the insurer improves its ability to predict its expected losses. Although most insurers collect in advance premiums that will be sufficient to pay all their expected losses, some rely at least in part on assessments levied on all insured after losses occur. Insurance, as risk treatment operation, is usually applied to financing risks of high severity and high frequency.

1.39 Management Concepts

Management may be defined as the process that achieves the most effective combination of available resources for implementing the objectives of a firm.

The fundamental functions performed by managers are the following.

Planning

Organizing

Staffing

Directing

Controlling

The above functions constitute particularly important factors for every firm whatever form, size and sector of activity. Moreover, these functions are incorporated in every activity of a firm.

Management is the structural element which achieves the efficient handling of any collective effort and creation of a suitable environment encouraging the cooperation and maximization of the performance of each part, in order to implement the desired results.

Factors of production such as labor, capital, and nature are static. These factors should be connected to form a system in such a way as to have the best possible use. The element by which this is achieved is management.

A person exercises management if the person is responsible for achieving results through others. Basically, management as a process includes all those functions that are required for the correct utilization and optimization of available resources. Managers perform part of planning, organizing, staffing, directing, and controlling which are the functions of management process. It is quite obvious that management covers all the activities of the firm. But the content of management should not be confused with the implementation of activities framing the operations of the firm. Of course, in practice there is no complete separation between the work of management and the executive work. But the content of the work of management differs from the content of executive activity.

Although management functions are not completely separated from each other but they can be considered specific elements of procedures arising in the implementation of management process. In the following subsections, the functions of management are described as independent and well separated from each other, for facilitating the understanding of each function. The order of presentation of management functions shows the physical order of these functions in practice or equivalently planning, organizing, staffing, directing, and controlling.

The function of planning includes the following activities.

Making predictions for the future behavior of all variable factors of the firm and its environment affecting the activities of the firm

Defining the objectives of the firm, which shaping the behavior of the firm

Identifying and analyzing of alternative ways of action

Decision making for the selection of optimal ways of action

Coordinating all the actions which must be done to achieve the objectives of the firm

It is necessary to emphasize that the function of planning is a very important function of management. The functions of organizing, staffing, directing, and controlling are based on the function of planning.

The use of informatics and operational research in the study of factors affecting the activity of a firm has substantially improved the function of planning and particularly the procedures of making predictions and decisions.

The function of organizing includes the following activities.

Dividing the work which is to be done in individual activities

Grouping these activities into separate sections

Assigning the corresponding authority to the persons responsible for these sections in order these persons to be able to carry out their tasks in the most efficient manner

It is obvious the function of organizing is not an end in itself but a means to achieve specific objectives. The more complex the objective is the more complex is the function of organizing and so complex relationships are needed to achieve this objective.

The function of staffing includes the care for continuous manning of the firm with appropriate people. The organizational structure determines the positions, or equivalently the duties and authority of each individual. The different positions should be covered by individuals with only criterion of selection the suitability of individuals for the requirements of the positions. The proper staffing substantially contributes to the successful implementation of the program and the profitability of the firm. For this reason, the issue of human factor is included in the main interests of the firm.

The function of directing concentrates on the guidance of human resources in such a way that every effort should be directed at achieving the objectives of the firm. The key element of directing is leadership, whereby the human resources will be directed towards meeting the objectives of the firm.

Basically, the function of directing includes the following activities.

Determining human needs, which as motives shape human behavior

Determining those needs, which may be activated and fulfilled into the firm, and thus those needs may act as motives of behavior in the firm

Investigating tools which can be used to satisfy needs of human resources, such as large salary, securing employment, ethical recognitions, and promotion opportunities

Concurrency of the *trend of satisfaction of human needs with the effort to carry out the objectives of the firm, so that the measure of the implementation of the objectives of the firm to provide the employees with the possibility of his personal needs*

Finding the appropriate system of guidance of *human resources or equivalently the democratic or authoritarian way to direct the human resources to objectives of the firm*

Controlling is the function which contributes to the achievement of objectives and the implementation of the program of the firm. Controlling involves the existence of objectives and program for the firm. Moreover, controlling completes the management process.

The function of controlling includes the following activities.

Measuring the results obtained, as production volume, the volume of material used, the degree of success of an effort of advertisement, the amount of earnings, and other results

Determining differences between planned and realized results

Identifying the causes, which created the differences between planned and realized results

Taking measures to correct the differences appeared

The correction of differences closes the circuit of management functions, because this correction requires the activation again of the other functions of management.

1.40 Management Approaches

The approaches which substantially contribute to the understanding and the effectiveness of management are the following.

Mathematical Approach

Decision Theory Approach

Empirical Approach

Human Behavior Approach

Social System Approach

Since management includes risk management then the present work can be considered as a part of the mathematical approach of management.

1.41 Mathematical Approach of Management

The mathematical approach of management makes use of mathematical models for the description of management problems and with the use of mathematical logic finds the possible solutions of these problems.

The theoretical basis of this approach, which is known as the approach of operational research, is the principle that the functions of planning, organizing, staffing, directing, and controlling are logical procedures and therefore these functions can be expressed in mathematical symbols and mathematical relationships.

The main topic of the mathematical approach of management is the possibility of quantitative formulation of management problems. This problem mainly arises in the human factor and the area of social relations, where the description of psychological and social phenomena is often impossible with concepts and even more with mathematical symbols and mathematical relationships. But this does not necessarily entail challenging the systematic effort of mathematical science to penetrate the disciplines of psychology, sociology, and in general the behavioral sciences offering thereby mathematics for solutions of psychological and social problems.

The application of mathematical science in many fields of human activity and various approaches of management means that the mathematical approach of management is a tool of solving management problems than such an approach.

In the relative literature the mathematical approach of management or the approach of operational research is identified as the quantitative approach of management.

1.42 Definition of Risk Management

Risk management may be defined as the systematic process of managing the risks threatening an organization in order to accomplish its goals in a way consistent with common interest, human protection, environmental factors and the law. It consists of the planning, organizing, directing, and controlling the safety operations of an organization with the aim of developing an efficient procedure that decreases the negative results of risks threatening that organization. Safety operations are usually called risk management operations. Risk measurement and risk treatment are generally considered as two particularly important risk management operations.

The formulation and investigation of stochastic models for risk measurement and risk treatment operations constitute two of the main purposes of the present work.

1.43 Risk Management and Unprecedented Uncertainty

Risk management is clearly still evolving as discipline. All of the current discussion, therefore, is extremely healthy, especially if it leads to practitioners from different disciplines talking and responding to one another. Risk management is and will remain a key ingredient of sound management in a world of unprecedented uncertainty.

1.44 Risk Management Philosophy

There are three factors that effectively define an organization's approach to managing risk, those relating to:

Structure—the nature of the organizational infrastructure

Strategy—the nature and combination of techniques used in risk management

Culture—the beliefs and values that influence the actions of individuals and groups, who are directly and indirectly responsible for risk management within the organization.

1.45 Relationship of Risk Management to Management

Although risk managers in discharge of their major responsibilities are often concerned with technical matters relating to risk analysis and treatment, they are also typically in charge of a department and/or an insurance subsidiary and frequently in contact with other functional areas in the organization. They often participate in the managerial decision process, particularly as it relates to risk management policy.

Thus, risk managers should be aware about the general principles of management. Among these basic principles that would seem particularly pertinent to the risk management function are the following.

Development of Objectives

Decision Making

Use of Quantitative Methods

Planning

Organizing

Staffing

Communication

Control

Management of Overseas Operations

Risk management as one of the specialties within the discipline of general management shares many of the characteristics of general management. However, the above operations are extremely important for implementing the goals of risk management.

1.46 Prerequisites of Applying Risk Management

The implementation of risk management requires organizing such that the assessment of risks is based on technocratic criteria, insurance knowledge, and systems avoiding the creation of stress to the economic strength and solvency of the firm. The following fundamental principles of insurance theory and practice are very important in applying risk management.

Fragmentation of risk and the subsequent dispersion of the economic impact of an adverse event

Securing the necessary reinsurance or coinsurance net of protection

Keeping overhead cost low and a reserve capable of responding to a catastrophic damage,

Elimination, as far as possible, of major risks through insurance combinations ensuring small insurance cost but also adequate protection

Empirical studies have found that a key factor in the effective implementation of risk management is the education of staff in the fundamental concepts and objectives of risk management.

1.47 Risk Management and Insurance

Historically, the risk management process has been closely linked with insurance. Many risk managers have come from insurance companies where they have acquired knowledge about insurer operations. Their role in the firm has been often that of an insurance buyer and their activities have centered around problems of risk transfer and those preventive activities that are closely related to insurance loss experience. Insurance experience is valuable since insurance companies are the primary example of businesses that undertake the risks of others. In spite of fairly widespread use of risk assumption techniques, insurance contracts remain the principal way of handling risk for many business firms. Thus any discussion of risk management must treat insurance at some length, including the principles surrounding its purchase.

1.48 Risk Management Objectives

Risk management has several fundamental goals. These objectives can be divided into the following two categories. The first category incorporates the objectives of risk management to be achieved prior to the occurrence of a risk. The second category incorporates the goals of risk management to be achieved after the occurrence of a risk.

The most fundamental objectives of risk management belonging to the first category are the following.

Economy

The economy objective means that the firm should prepare for potential losses in the most economical way. This involves a financial analysis of safety program expenses, insurance premiums, and the costs associated with the different techniques for handling losses.

Reduction of Anxiety

This objective is more complicated. Certain loss exposures can cause greater worry and fear for the risk manager, key executives, and stockholders than other exposures. However, the risk manager wants to reduce the anxiety and fear associated with all loss exposures.

Meeting Externally Imposed Obligations

This objective is to meet any external imposed obligations. This means the firm must meet certain obligations on it by outsiders. For example, government regulations may require a firm to install safety devices to protect workers from harm. Similarly, a firm's creditors may require that property pledged as collateral for a loan must be insured. The risk manager must see that these externally imposed obligations are met.

The most fundamental objectives of risk management belonging to the second category are the following.

Survival

The first and most important risk management objective after a loss occurs is survival of the firm. Survival means that after a loss occurs, the firm can at least resume partial operation within some reasonable time period if it chooses to do so.

Continuity of Operations

The second post-loss objective is to continue to operating. For some firms, the ability to continue operating after a severe loss is an extremely important objective. This is particularly true of certain firms, such as a public utility firm, which is obligated to provide continuous service. But it is also extremely important for those firms that may lose some or all of their customers to competitors if they cannot operate after the occurrence of a loss. This would include banks, bakeries, dairy farms, and similar firms.

Stability of Earnings

Stability of earnings is the third post-loss objective. The firm wants to maintain its earnings per share after a loss occurs. This objective is closely related to the objective of continued operations. Earnings per share can be maintained if the firm continues to operate. However, there may be substantial costs involved in achieving this goal, such as operating in another location, and perfect stability of earnings may not be attained.

Growth

The fourth post-loss objective is continued growth of the firm. A firm may grow by developing new products and markets or by acquisitions and mergers. The risk manager must consider the impact that a loss will have on the firm's ability to grow.

Social Responsibility

Finally, the objective of social responsibility is to reduce the impact that a loss has on other persons and society in general. A severe loss by a firm can adversely affect employees, customers, suppliers, creditors, taxpayers, and the community in general. For example, a severe loss that requires shutting down a plant in a small community for an extended period can lead to depressed business conditions and substantial unemployment in the community. The risk manager must therefore be concerned about the social responsibility of the firm to the community after a loss occurs.

1.49 Risk Management and Human Needs

Man is motivated to fulfill his needs, where the need is the avoidance of an undesirable condition or the pursuit of some desired condition. The source of man's motivation to fill his perceived needs has been a major concern of psychological theory and research. Maslow categorizes and ranks basic sets of human needs into a conceptual hierarchy. This hierarchy is important to analysis of risk because it provides a basis for understanding how man values possible gains and unwanted consequences. This, coupled with the idea that healthy versus abnormal people are used to study behavior, makes the results more palatable for understanding normal, individual, societal risk behavior. Maslow made the primary breakdown as follows in hierarchy of needs.

Physiological Needs

The physiological category refers to food, warmth, shelter, elimination, water, sleep, sexual fulfillment, and other bodily needs.

Safety Needs

The safety needs include actual physical safety, as well as a feeling of being safe from physical and emotional injury; therefore, a feeling of emotional security as well as a feeling of freedom from illness.

Need for Belongingness and Love

Whereas physiological and safety needs are centered round the individual's own person, the need for belongingness and love represents the first social need. It is the need to feel a part of a group or the need to belong to and with someone else. It implies the needs to both give and receive love.

Need for Esteem

The need for esteem is based on the belief that a person has a fundamental requirement for self-respect and the esteem of others, except in extreme pathology. The need for esteem is divided into two subsets. First, there is the need for feeling a personal worth, adequacy, and competence. Second, there is the need for respect, admiration, recognition, and status in the eyes of others.

Need for Self-actualization

Self-actualization is a more difficult concept to describe. It is the process whereby one realizes the real self and works toward the expression of the self by becoming what one is capable of becoming. Thus the need for self-actualization sets into motion the process of making actual a person's perception of his "self".

The above hierarchy emphasizes the fundamental point that until one need is fulfilled, a person's behavior is not motivated by the next, higher level, need. For example, a person whose physiological needs are not taken care of, is not concerned with his safety. And until his physiological and safety needs are not fulfilled, he is not particularly interested in fulfilling his need for love. Moreover, the above hierarchy, the definition and the main goals of risk management discipline make quite clear that this discipline substantially contribute to the satisfaction of human needs.

1.50 Safety Needs-Workplace—Risk Taking

In the workplace, safety needs manifest themselves in the form of confidence for this post. Safety needs, in this sense, arise when people are under conditions of dependency on other people or organizations and they are under the fear of arbitrary dismissal. The feeling that people have the fairest possible treatment meets the safety needs in the workplace. When people feel that they are treated properly, they tend to take risks and they become more creative.

Almost all people working in commercial and industrial organizations are in relationship of partial dependency. This relationship gives particular attention to safety needs. The feeling of insecurity stems from arbitrary administrative behavior that creates uncertainty which is strong incentive in shaping safety needs at each level of the hierarchy of personnel. This factor explains at a large part the growing trend to syndicalism.

Major objective of risk management is to meet human safety needs in the workplace.

1.51 Spiritual Growth and Future Needs

Primitive man seeking to meet the needs of the present he did not see the needs of the future, he had no fear of this and he did not make predictions for the future. Since man spiritually grows, creates civilization, he can maintain and compare information, draw conclusions from the past for the future, he is more afraid of the future and he continuously thinks to find protection against eventual adverse conditions of the future, he feels the intensity of the future needs, and avoids consuming goods, saves them for his future needs.

The importance of the role of risk management in modern society is revealed by the powerful bond between the spiritual growth of man and the vision of his future.

1.52 Social Progress and Proactive Risk Management

Social progress is mainly based on the following factors.

Advancement of Human Intellectual Characteristics

Improvement of the Material Status of Man

Subordination of Individual Human Instincts to the Ideals of Human Life in Society

The particular importance of the proactive component of risk management substantially supports the factors of social progress.

1.53 Cost of Risk

Each of the insurable risks to which an organization is exposed should be seen as a cost composed of interrelated factors. This cost is called cost of risk and incorporates the following factors.

Cost of Loss Prevention

Insurance Premiums

Losses Sustained Net of Indemnities from Insurers and Third Parties

Expenses of Administration

The cost of risk is the sum of the above four factors, and the objective of risk management should be to minimize this cost by changing the amenable factors as appropriate. The cost of risk is mathematical in principle, but since some of its factors cannot be accurately forecasted or measured, a precise net total cannot be calculated. However, since the factors can be estimated, it provides practical direction for the efforts of the risk managers. Today, in many organizations, the cost of risk is used as both an internal and external benchmark for performance measurement.

1.54 Data-Information-Knowledge

Data are series of unconnected events and observations which can be converted to information if we analyze, summarize or organize them in some other way. It is, therefore, required work to convert data to information. Information is more valuable than data. Data are transformed into information according to a sense for specific purposes. Information has been overtaken land, labor, and capital and is the most important inflow in modern production systems. Information reduces the needs for labor, land, and capital. It creates whole new industries. It is separately present in the market and is also the raw material of the fastest growing sector of the industry of knowledge. The labor market is now dominated by information operators paid since they posses necessary information to implement objectives.

In turn, the sets of information can be processed to form a coherent system of knowledge.

Knowledge consists of organized sets of information which form the basis of theoretical conceptions and evaluations.

It is generally accepted that acquisition, creation, and use of information greatly assist risk managers in the implementation of the tactical and strategic objectives of an organization.

1.55 Information Operators

The labor market is currently dominated by the following kinds of information operators.

Information Organizers

This kind of information operators deals with the appropriate organizing of information involved in modern production systems. The senior managers constitute this kind of information operators.

Information Transmitters

This kind of information operators deals with the transmission of information. Secretaries, journalists and other employees in mass media, and educational staff of any category are information transmitters.

Information Administrators

This kind of information operators deals with the storage and retrieval of information. Librarians and computer programmers are information administrators.

Information Creators

The creation of new information is the purpose of this kind of information operators. Researchers of any category are information creators.

Information Users

This kind of information operators deals with the use of information for solving specific problems. Lawyers, doctors, and consultants of any category are information users.

Information Receivers

This kind of information operators is engaged in the acquisition of information. Students of any category are receivers of information.

The chief risk officers and senior risk managers work as organizers, transmitters, administrators, creators, users, and receivers of information in modern complex organizations.

1.56 Disciplines Contributed to Risk and Risk Management

The perception of the complex concept of risk and the essence of risk management are substantially facilitated by making use of data, information, knowledge, and theories from various disciplines. Such a use is an extremely difficult process. The following sections incorporate several extracts from some disciplines. These extracts contribute to a very good consideration of the possibilities of risk management. Such extracts are provided by the following disciplines.

Anthropology

History

Psychology

Neuroscience

Philosophy

Informatics

Literature

Management

Cindynics

Systemics

Mathematics

1.57 Risk Avoidance and Evolution of Memory

Antonio Damasio believes that memory requires the mobilization of many brain systems that work together seamlessly to different levels of neural organization. He argues that memory was evolved to help us to avoid risks, based on our experience and not just to reminisce. And that is why we tend to remember what they seem novel or important, while the rest are indiscriminately forgotten. This procedure is considered necessary, since remembrance of everything would cause an explosive overload of the brain circuits.

Antonio Damasio
Descartes' Error: Emotion, Reason, and the Human Brain
Putnam Publishing, 1994

1.58 Imaginary Risk and Brainwashing

Man is not only able to predict real risks in the future, he may even be convinced or brainwashed by his leaders that he faces risks and when there are not actually risks. Most modern wars are prepared by systematic propaganda of this type. People are convinced by their leaders that there is a risk to be attacked and destroyed, so they react with hatred against nations which threaten them. Often there is no threat. Especially since the French Revolution that large armies of citizens appeared rather than relatively small bodies of professional soldiers, it is not easy to the leader of a nation to tell people to kill and to be killed because the industry wants cheaper raw materials, or cheaper labor hands or new markets. Only a minority would agree to participate in a war with such purposes. If on the other hand, a government can make people to believe that they are threatened then the normal biological reaction to the threat has been put into operation. Moreover, these predictions of the threat from the outside bring often themselves the result: the aggressor nation with its martial preparations forces the nation that is going to suffer the attack to make respective martial preparations "demonstrating" so the alleged threat.

Erich Fromm
The Anatomy of Human Destructiveness
Holt, Rinehart & Winston, 1973

1.59 Specific Human Risk

Man was born as a whim of nature, being part of nature, and overtaking it. He must find principles for action and decision making to replace the principles of instinct. He must have a framework of orientation that allows him to organize a consistent picture of the world as an assumption for consistent actions. He must fight not only the risk of dying, starving, and hurting but also another risk which is specifically human: the risk of ending up mad. In other words, he must protect himself from the risk of losing his life, but also the risk of losing his mind.

Erich Fromm
The Revolution of Hope: Toward a Humanized Technology
Harper & Row, 1968

1.60 Risk Acceptance and Results of Calculations

The acceptance of a risk is not always the result of calculations—far from it. In the economic sector, Keynes observed. "If human nature would not love risk, if the construction of a factory or a railway, the exploitation of a mine or a farm would not give any other satisfaction than that offered by profit, then such a scale of investments would not exist".

Ivar Ekeland
The Broken Dice and Other Mathematical Tales of Chance
University of Chicago Press, 1993

1.61 Increasing Risk and Human Relations

We must always be conscious that the negative factor may play a positive role. We often mention the famous verse of Hoelderlin: "But where the danger is, also grows the saving power". We are in such a situation. Hopefully the increasing risk will not keep the pace with the increasing probabilities for salvation with the religious sense, but with the improvement of human relations, which everyone, in the depth of his self, expects.

Hoelderlin

1.62 Real Risk and Extraordinary Energy

A real risk tends to mobilize extraordinary energy for its treatment, often to the extent that the person concerned could never put to its mind that it has such a great physical strength, skill and endurance. However this extraordinary energy is mobilized only when the entire body faces a real risk and for neurophysiologic reasons, risks dreaming of the person do not irritate the body in this way, but only lead to fear and anxiety.

Erich Fromm
The Anatomy of Human Destructiveness
Holt, Rinehart & Winston, 1973

1.63 Risk of Excessive Speed of Development

Our civilization is a victim of speed. Awareness of the mad pace, the risk of the excessive speed of development is urgent. We have to brake, to slow, for being able to look at a different future. It is now necessary to think and plan an international regulation of growth and economic competition and to promote the creation of a charter which will include the right to human time.

Edgar Morin
Anne Brigitte Kern
Terre Patrie
Editions du Seuil, 1993

1.64 Overthrow of Primitive Human Nature

In selective adaptation in front of the risks of the Stone Age, human society exceeded or subjugated the primitive tendencies as selfishness, criminal sexuality, dominance, and fierce competition. Human society put the harmony and the cooperation in place of conflict, solidarity over sex, and ethics over power. In these early days human society conducted the biggest overhaul in its history, the overthrow of the primitive human nature, thus ensuring a future full of evolution for the species.

Erich Fromm
The Anatomy of Human Destructiveness
Holt, Rinehart & Winston, 1973

1.65 Risk and Need

A heart full of courage and cheerfulness needs a little danger from time to time, or the world gets unbearable.

Friedrich Nietzsche
R. Flesch
The New Book of Unusual Quotations
Harper & Row, Publishers, Inc., 1966

1.66 Playing Man

Man is neither entirely preform of the genetic code or simple result of the action of his environment. His behavior is not the set of his absolute or dependent reflective nor an expression of absolute freedom. Man is neither absolute subject nor absolute object. Maybe that is why it not easy a definition of man. However, man has a proper name: Homo, and quite a few adjectives: sapiens (wise), laborans (working), faber (creator), loquens (talking), oeconomicus (economic), even ludens (playing). Attention to last adjective. Sometimes man plays with fire.

Johan Huizinga
Homo Ludens: A Study of the Play—Element in Culture
Beacon Press, 1955

1.67 Developing Courage and Challenging Adversity

You do not develop courage by being happy in your relationships every day. You develop it by surviving difficult times and challenging adversity.

Epicurus

1.68 Adversities and Evolution

For the modern observer, who looks at the life story through his view of many millions years, some meaning is obvious. He sees that, with the struggle against adverse conditions, each generation shapes the forms that its offspring will have. The adversities and struggles are at the bottom of the pyramid of evolution. Without adversities there is no pressure and without pressure there is no change. These circumstances, so persistent on each individual, create currents needed for life to evolve and reach from the most simple to the most complex. Finally, man stands on the earth, most perfect of all the other creatures that inhabit it. Smart and aware of own behalf, he alone of all the other creatures has the curiosity to ask himself about the way of his creation and the forces which created him. Moreover, having as a guide his scientific knowledge understands that he was created by all those who lived before him, fighting to overcome the difficulties that they were finding ahead.

Robert Jastrow
Until the Sun Dies
Warner Books, 1977

1.69 Risk in Society

The notion of risk becomes central in a society which is taking leave of the past, of traditional ways of doing things, and which is opening itself up to a problematic future. This statement applies just as much as to institutionalized risk environment as to other areas. Insurance is one of the core elements of the economic order of the modern world it is part of a more general phenomenon concerned with the control of time. The 'openness' of things to come expresses the malleability of the social world and the capability of human beings to shape the physical settings of our existence. While the future is recognized to be intrinsically unknowable, and as it is increasingly severed from the past, that future becomes a new terrain, a territory of counterfactual possibility.

Anthony Giddens
Modernity and Self-Identity
Polity Press, 1991

1.70 Risky Hunan Confusion and Impact of Computer Revolution

Even the most optimistic fan of the human race will have to concede that our world has reached a very risky stage of confusion and man, helpless, is rather unlikely to accomplish many things in an attempt to rectify the situation. Now many people believe that the more man left alone to manage their affairs, so it is compounded the confusion—and the only solution would be to shake everything in the air with hydrogen bombs-which unfortunately we cannot exclude it. Awareness of our hopeless position in a world unimaginably complex and loaded with information there that does not get another, it will become deeper in troubled and restless years for decades, when we feel in all its breath and intensity the impact of revolution of computers. Under these circumstances, the temptation to turn to computers for support will be invincible. Of course, once we succumb to it, everything will change. Man, only and undisputed master of this planet for centuries, will no longer treat the universe alone. Other intelligent beings, in the beginning equal in authority to him and later senior, will stand by his side.

Christopher Evans
The Mighty Micro
Victor Gollancz Ltd, 1979

1.71 Risk and Possession of Software

Possession of software is not risky, but possession of large amounts of software is risky. The barnacle has trivial software—and that software has been devoted to the basic needs of her digestive and reproductive system and therefore barnacle has not software left to put her into trouble. However, man has large amounts of software that a significant part of it is dedicated to ensuring the survival in a world full of beings predators and prey, aggressors and victims. Unfortunately, for the most part, this software is instinctual and tends to be exhausted in ruthless and deeply selfish aggression. The remaining part of the software, which is non-instinctual or acquired, tends to rotate in a different direction. The balance, however, leans towards the instinctual part as a result, while man has gained great technological superiority in this world, when he is threatened, he is able to turn with tremendous comfort to the programs that he inherited from his distant past of the jungle and caves.

Christopher Evans
The Mighty Micro
Victor Gollancz Ltd, 1979

1.72 Risk of Atomic War and Genetic Material

DNA, the genetic material, is the most wonderful thing in the world, and it is protected by the nature very carefully. Mankind went through epidemics, pests and all kinds of tribulations, but nature kept this material intact, because whole life depends on it. Today, for the first time in history, man has discovered the means of destructing the genetic material. High radioactivity achieves the destruction of the genetic material. This destruction is the big change in the nature of atomic war. In the previous wars DNA was not under direct threat and humanity could continue to live. There may be survivors after the atomic war, but they will be unable to produce healthy offspring. The offspring of the survivors will suffer from abnormalities, deformities and diseases, which will make life unbearable, with no way back to our time.

Albert Szent-Györgyi
The Crazy Ape
Philosophical Library, 1970

1.73 Risks of Savanna

The gradual replacement of mild, protective, and capable of providing food, forest by relentless aggressive savanna spurs and drives the process of anthropogenesis. Savanna creates the conditions of faceted use of two-legged, two-handed, and cerebral skills, starting from the needs and the risks it creates. The new ecosystem, indeed, involves its coercive powers, guidelines and risks, which are stimuli to develop all sorts of skills that already exist in the ancestor of forest, which, relative of the chimpanzee, gifted with an agile brain, with a piercing glance and an omnivorous appetite, is able to convert a branch to a bat and a pebble in a projectile and downs collectively, small mammals.

Edgar Morin
Le Paradigme Perdu: La Nature Humaine
Editions du Seuil, 1973

1.74 Inability of Treating New Risks

Let us imagine the invasion of "conquistadores" to the empire of Incas. The society of the Indians knew the acceptance of certain risks. It was an agricultural society, so it was familiar with the risks of agricultural life. It was also an empire that had been established after conquests, so it knew and the military risks. But it did not treat the new risk posed by these bearded beings with armors, which were riding unfamiliar animals, holding sticks which were throwing lighting, and provoked the usual laws of life. The reasons of this collapse have disappeared along with Atahualpa and the millions of his subjects, but it seems likely to be associated with some psychological inability of accepting certain risks. It is preferable to surrender than to fight with an opponent whose strength cannot be measured.

Ivar Ekeland
The Broken Dice and Other Tales of Chance
University of Chicago Press, 1993

1.75 Boldness and Risk of Death

Courage is the ability of someone to resist the temptation to risk hope and faith—and thus destroy them—transforming them into hollow optimism or irrational belief. Courage is the ability to say "no" when the world wants to hear "yes". But courage cannot be fully understood if we do not consider it from another point of view: boldness. The bold person is not afraid of threats, even his death. However, often, the word "bold" covers various, totally different attitudes. I mention the most important: First, a person can be bold because the person does not care to live or not. Life is not worth a lot of things about the person so the person is shown bold in front of risk of dying. But while the person is not afraid of death, he may be afraid of life. His boldness is based on lack of love for life. Usually the person is not bold when he is not in front of risk of dying. In fact the person often searches risky situations trying to avoid the fear of life for himself and for the people.

Erich Fromm
The Revolution of Hope: Toward a Humanized Technology
Harper & Row, 1968

1.76 Specific Boldness and Risk of Isolation

A specific kind of boldness is the boldness of a person which lives in a symbiotic allegiance to an idol, whether it is a person or institution, or idea. The commands of the idol are sacred. These commands are more compulsory than the commands of the survival of the body. If he could disobey or challenge the commands of the idol, he would face the risk of losing his identification with the idol. This means that the person would face the risk to be completed isolated, so the brink of madness. He wants to die because he is afraid to expose himself to this risk.

Erich From
The Revolution of Hope: Toward a Humanized Technology
Harper & Row, 1968

1.77 Fear of Risk Taking

To try to eliminate risk in business enterprise is futile. Risk is inherent in the commitment of present resources to future expectations: Indeed, economic progress can be defined as the ability to take greater risks: The attempt to eliminate risks, even the attempt to minimize them, can only make them irrational and unbearable: It can only result in the greatest risk of all: rigidity:

David B. Hertz
Howard Thomas
Risk Analysis and its Applications
John Wiley and Sons, 1983

1.78 Inevitability of Thinking in Terms of Risk

Thinking in terms of risk becomes more or less inevitable and most people will be conscious also of the risks of refusing to think in this way, even if they may choose to ignore those risks. In the charged reflexive settings of high modernity, living on 'automatic pilot' becomes more and more difficult to do, and it becomes less and less to protect any lifestyle, no matter how firmly pre-established, from the generalized risk climate.

Anthony Giddens
Modernity and Self-Identity
Polity Press, 1991

1.79 Risk and Time

Risk and time are opposite sides of the same coin, for if there were no tomorrow there would be no risk. Time transforms risk, and the nature of risk is shaped by the time horizon: the future is the playing field.

Peter Bernstein
Against the Gods
The Remarkable Story of Risk
John Wiley & Sons, Inc., 1996

1.80 Mastery of Risk

The revolutionary idea that defines the boundary between modern times and the past is the mastery of risk: the notion that the future is more than a whim of the gods and that men and women are not passive before nature. Until human beings discovered a way across that boundary, the future was a mirror of the past or the murky domain of oracles and soothsayers who held a monopoly over knowledge of anticipated events.

Peter Bernstein
Against the Gods
The Remarkable Story of Risk
John Wiley & Sons, Inc., 1996

1.81 Conversion of Risk Taking

This book tells the story of a group of thinkers whose remarkable vision revealed how to put the future at the service of the present. By showing the world how to understand risk, measure it, and weigh its consequences, they converted risk taking into one of the prime catalysts that drives modern western society. Like Prometheus, they defied the gods and probed the darkness in search of the light that converted the future from an enemy into an opportunity. The transformation in attitudes toward risk management unleashed by their achievements has channeled the human passion for games and wagering into economic growth, improved quality of life, and technological progress.

Peter Bernstein
Against the Gods
The Remarkable Story of Risk
John Wiley & Sons, Inc., 1996

1.82 Risks and Globalization

Risks which are currently preoccupying the international community are not divisible into neat compartments or equally distributed. Globalization is seen by some as representing the possibility of eradicating disease, increasing access to markets, utilizing productive technology and modern management methods for the benefit of all. What is less frequently mentioned are the risks which are inherent in the process of globalization, which go beyond the immediate set of ecological risks. There is a real risk that some countries could become marginalized from the world economy, and that this would cause inequality within nations, unless we manage the process of globalization in a way which ensures that its fruits are equally distributed.

David Denney
Risk and Society
SAGE Publications, 2005

1.83 Nature of Man

Man is basically a risk aversive animal who usually seeks to avoid or minimize risk. However to achieve some perceived benefit, man will undertake to increase his risk. For example, men risk their lives with near certainty of premature death in war or peace to achieve the perceived benefits of freedom. This subjective, qualitative balancing of risks and benefits is an innate ability of man as a rational animal and serves as a model for analytic processes as undertaken here and elsewhere. Although aversive to risk, man faces risks he cannot control, including the certainty of death. He evidently has built in physiological and emotional blocks that permit him to ignore risks he can do nothing about and to go on living his life in a pragmatic manner. Man also rationalizes unpleasantness, including risks, by blanking them out: "out of sight, out of mind".

William D. Rowe
An Anatomy of Risk
Robert E. Krieger Publishing Company, 1988

1.84 Risk Taking and Values

The accident rate and the incidence of unhealthy habits depend on people's orientation towards their future. The more they expect from it, the more careful they will be with life and limb. If their expectations are low, they will try to find more immediate gratification of their desires, and do so at a greater risk of jeopardizing their lives. The extent of risk taking with respect to safety and health in a given society, therefore, ultimately depends on values that prevail in that society, and not on the available technology.

Gerald J.S. Wilde
Target Risk
PDE Publications, 1994

1.85 Risk and Numbers

Modern methods of dealing with the unknown start with measurement, with odds and probabilities. Without numbers, there are no odds and no probabilities; without odds and probabilities, the only way to deal with risk is to appeal to the gods and the fates. Without numbers, risk is wholly a matter of gut.

Peter Bernstein
Against the Gods
The Remarkable Story
John Wiley & Sons, Inc., 1996

1.86 Risks and Human Activities

Risks are ultimately caused by human demands and needs that generate human activities. Examples of such activities are developing and operating a chemical plant, living below sea level, drinking alcohol and traveling by car. Such activities can lead to damage or loss involving human health, the environment or goods. The aim of human activities is to produce benefits, but the inevitable side effect is the creation of risks. This is why risks cannot be viewed separately from benefits.

Passchier, W.
Reij, W.
Risk is More Than Just a Number
RISK: Health, Safety & Environment, Vol. 8, 1997

1.87 Risky Shift

Risk shift is defined as the tendency of certain groups to become more extreme or take riskier positions in their judgements than they would, acting as individuals.

William D. Rowe
An Anatomy of Risk
Robert E. Krieger Publishing Company, 1988

1.88 Risk Acceptance

The willingness of an individual, group, or society to accept a specific level of risk to obtain some gain or benefit is defined as risk acceptance.

William D. Rowe
An Anatomy of Risk
Robert E. Krieger Publishing Company, 1988

1.89 Risk as a Stimulus

An extremely important challenge for man is to learn how to live with uncertainty so that risk can be an acceptable stimulus, rather an unacceptable threat.

Felix H. Kloman
Rethinking Risk Management
The Geneva Papers on Risk and Insurance, 17, 1992, 299–313

1.90 Risk as a Choice Rather Than a Fate

The word "risk" derives from the early Italian "risicare", which means "to dare". In this sense, risk is a choice rather than a fate. The actions we dare to take, which depend on how free we are to make choices, are what the story of risk is all about. And that story helps define what it means to be human being.

Peter Bernstein
Against the Gods
The Remarkable Story of Risk
John Wiley & Sons, Inc., 1996

1.91 Essence of Risk Management

The essence of risk management lies in maximizing the areas where we have some control over the outcome while minimizing the areas where we have absolutely no control over the outcome and the linkage between effect and cause is hidden from us.

Peter Bernstein
Against the Gods
The Remarkable Story of Risk
John Wiley & Sons, Inc., 1996

1.92 Real Objective of Risk Management

Perhaps the real objective of risk management is to reduce fear of the unknown and the unexpected, and to create confidence in the future. It is certainly not simply the technical understanding of risk nor making financial provisions for alleviating the pain of loss. To carry this theme further, could it have been some inchoate fear of the future that spawned the incredible materialism and greed that have been the symbols of the past decade? Are we so afraid of the future that, in compensation, we devour the present, and, perhaps ironically, make a secure future less possible? Considering the broad set of freedom from fear, we can conclude that there may be a more significant role for risk management, to restore our faith in ourselves and our futures, and to reestablish our confidence in our abilities to survive and prosper. This is why rethinking risk management is so important today, in a period in which we are beset by uncertainties. If risk management can contribute to the larger goal of creating new confidence, reducing fear, it will be important to all of us both individually and in the organizations and political structures in which participate.

Felix Kloman
Rethinking Risk Management
The Geneva Papers on Risk and Insurance, 17, 1992, 299–313

1.93 Autopoiesis and Risk Management

To be alive, an entity must be autopoietic, that is, it must actively maintain itself against the mischief of the world. Life responds to disturbance, using matter and energy to stay intact. An organism constantly exchanges its parts, replacing its component chemicals without ever losing its identity. This modulating, holistic phenomenon of autopoiesis, of active self maintenance, is at the basis of all known life; all cells react to "external perturbations in order to preserve key aspects of their identity within their boundaries".

Felix Kloman
Autopoiesis
Risk Management Reports, September 1996, Volume 23, No. 9

1.94 Uncertainty Principle of Risk Management

You will never know the losses avoided because of good risk management, but you will eventually know the consequences of inadequate risk management.

Chuck Marshall
Risk Management Reports
August 1997, Volume 24, No. 8

1.95 Risk Management and Probability Theory

As the years passed, mathematicians transformed probability theory from a gamblers' toy into a powerful instrument for organizing, interpreting, and applying information. As one ingenious idea was piled on top of another, quantitative techniques of risk management emerged that have helped trigger the tempo of modern times.

Peter Bernstein
Against the Gods
The Remarkable Story of Risk
John Wiley & Sons, Inc., 1996

1.96 Fundamental Factors of Risk Management Evolution

The following five factors could have significant effect on the continued development of risk management as an important organizational discipline.

Risk Communication

Balkanization of Risk Management Practice

Short term versus Long term Thinking

Effect and Threat of Catastrophes

Role of the Insurance Industry

Felix Kloman
Rethinking Risk Management
The Geneva Papers on Risk and Insurance, Vol. 17, 1992, 299–313

1.97 Structural Elements of Risk Management

The structural elements of risk management are mainly provided by the following scientific disciplines.

General Management Theory
Economics
Systemics
Operational Research
Probability Theory
Decision Theory
Informatics
Behavioral Science

Felix Kloman
Rethinking Risk Management
The Geneva Papers on Risk and Insurance, Vol. 17, 1992, 299–313

1.98 Risk Management and Quality Control

Quality control is risk management in a particular area, and it would fit into a risk management department that had appropriate status in the corporate organization.

Douglas Barlow
The Evolution of Risk Management
Risk Management, April 1993

1.99 Risk Manager Activities

The risk manager cannot be expected to personally minimize the adverse effects of losses, but rather to act as an adviser and coordinator. In short, the risk manager should be the catalyst to consolidate the efforts of all managers who control the organization. Making risk management happen in an organization requires more than just knowing risk management. It requires ingenuity and creative thinking.

Keith R. Gibson
Making Risk Management Happen in Your Organization
Risk Management, April 1991, Vol. 38, No. 4

1.100 Risk Management as Necessity

Risk exists whenever the future is unknown. Because the adverse effects of risk have plagued mankind since the beginning of time, individuals, groups, and societies have developed various methods for managing risk. Since no one knows the future exactly, everyone is a risk manager not by choice, but by sheer necessity:

C. Arthur Williams, Jr.
Richard M. Heins
Risk Management and Insurance
McGraw-Hill International Editions, 1985

1.101 Risk Management and Decision Making

The ability to define what may happen in the future and to choose among alternatives lies at the heart of contemporary societies. Risk management guides us over a vast range of decision making, from allocating wealth to safeguarding public health, from waging war to planning a family, from paying insurance premiums to wearing a seatbelt, from planting corn to marketing cornflakes.

Peter Bernstein
Against the Gods
The Remarkable Story of Risk
John Wiley & Sons, Inc., 1996

1.102 Risk Management and Systemics

System, in the general sense of the term, is any kind of organization that integrates space time and permits movement and development in the environment. Examples of systems are a cell, a business organization, a society, a city, or a country. The present section concentrates on the fundamental concepts of systems and the relationships of these concepts with the important concept of risk.

1.102.1 Environment of System

Various flows—information, obligations, products, risks—exist between any system and its environment. The system is influenced by and acts on its surroundings. What are the interactions? What is their value? Where are the risks?

1.102.2 Aim of System

Any system is made up of interacting elements oriented to perform system's aim. Each element depends on its contribution to the whole. Is each element well positioned in the whole? Does each element contribute to the risk? If yes, does it integrate approaches to measure, control and reduce the risk in its contribution to the whole?

1.102.3 Transformation

Each system in action puts its resources to work to produce foreseeable changes on the incoming flows: This is the mission of the system. The risks are part of it.

1.102.4 Products of System

Products are the result of this transformation and are identified by the divergence between the condition at input and after output. Is this change observable? Does this increase or decrease risks? In particular, are the product's risks under sufficient control, and acceptable by the system's surroundings?

1.102.5 Driving Mechanism of System

The system's driving mechanism consists of the intervention of the elements—variables of action—allowing the guiding of the device towards the attainment of the objectives. Risk management is a matter of risk avoidance, reduction, transfer and financing, but does a driving mechanism for this process exists? Is it active regarding risks? What does the guidance mechanism tell us about risks?

1.102.6 Feedback Information

A system also has a tool that provides information on the system's condition. This information has to be transmitted to the driving mechanism. Are there detectors? Do they take risks into consideration such as risks of loss, but also loss control, measuring errors and processing errors?

1.102.7 Regulation

A system is also endowed with an interfering mechanism, aimed at stabilizing the actualization of objectives on the defined level, and guided by feedback. Is that feedback information utilized for consequent action? Is the system itself under control, intervening at the appropriate time, at an adequate level and through appropriate means on the components or interactions responsible for any dysfunction?

1.102.8 Anticipation\Resilience

A system must anticipate changes that will occur in the surrounding environment. As a preventive measure, the system activates its adaptation reactions, but is it functioning? Since anticipation is not always possible, is the system appropriately trained to launch the reaction even in unanticipated circumstances? This adaptiveness and ability for resistance is called "resilience". It is in this context that the philosophy of risk management has to be proved.

Francois Settembrino
Risk Management in Enterprise: A Systemic Approach
Risk Management, August 1994, Vol. 41, No. 8

1.103 Combining Elements of Risk Management and Resilience

Resilience is the ability of a system and its component parts to anticipate, absorb, accommodate, or recover from the effects of a shock or stress in a timely and efficient manner. Moreover, resilience a concept concerned fundamentally with how a system, community or individual can deal with disturbance, surprise, and charge, is framing current thinking about sustainable futures in an environment of growing risk and uncertainty. Resilience has emerged as a fusion of ideas from multiple disciplinary traditions including ecosystem stability, psychology, the behavioural sciences and disaster risk reduction. Its recent appropriation by bilateral and multilateral donor organizations is one example of how resilience is evolving from theory into policy and practice. This appropriation has been driven by the need to identify a broad-based discourse and set of guiding principles to protect development advances from multiple shocks and stresses. Consequently, 'resilience' is an agenda shared by those concerned with financial, political, disaster, conflict and climate threats to development. Being a fusion of ideas and bridging many areas of development policy and practice, resilience poses particular challenges for programming

Tom Mitchell
Katie Harris
Resilience: A Risk Management Approach
Background Note
January 2012
Overseas Development Institute

1.104 Ascendancy of Risk Management

Managing risk is not confined to the work of professional risk analysts and managers. It has become an operational imperative for individual professionals, organizations and governments. Risk assessment and management constitutes a way of seeing the social world, which appears to be focused and managerially controllable as risks become more elusive and complex. Consumers of services expect risks to be identified and eliminated.

David Denney
Risk and Society
SAGE Publications, 2005

1.105 Risk Management and Crisis Management

Corporate leaders recognize that over the long term the only alternative to risk management is crisis management, and crisis management is much more expensive, time consuming and painful.

Chief Risk Officer of Fidelity Investments
Risk Mark
Famous Quotes

1.106 Risk Management and Cindynics

Cindynics is the new science of risks. As a subject of research, cindynics is developing rapidly with the increasing number of major risks and the increasing importance of risk management in modern complex organizations. The structural elements of the conceptual framework of cindynics are the following. A set of axioms on complex systems, a set of axioms for rationality operating in complex combination of networks, a set of axioms specific of cindynics, a concept of hyperspace for description of cindynic problems, a concept of cindynic situation, an operator of transformation of cindynic situations, a concept of cindynic dissonance, and a concept of event with a multidimensional structure. The international academic community of risk management recognizes and supports the role of the conceptual framework of cindynics to the development of various new research activities in the discipline of risk management.

Georges Kervern
Latest Advances in Cindynics
Economica, Paris, 1994

1.107 Risk Management and Stochastic Models

The use of mathematics in the investigation of real systems has become wide spread in recent times. This is partly due to the use of the scientific approach to problem solving and the increasing computational power of computers and computing methodology, both of which have made many more interesting problems amenable to a mathematical solution. These very significant factors have made possible the description, analysis, and solution with the use of mathematics of a very wide variety and exceptional importance of real problems arising in many practical disciplines.

The steps involved in using mathematics in the investigation of real world systems are problem formulation, mathematical description, mathematical analysis, and interpretation of analysis to obtain a mathematical solution. The most crucial and important step is the satisfactory translation of the system from the real world into a mathematical description. Once this is done, standard methods of mathematical analysis can be used to obtain a solution of the problem. The mathematical description is called a mathematical model and the sequence of activities for obtaining it is called mathematical modeling. While mathematical modeling has had a central consideration in economics, management, operational research, and physics for many years, it now appears not only to have achieved a similar status in such diverse fields as software, electronics, mechanics and other domains of engineering, but also in biological and psychological sciences as well. Several reasons were advanced to explain this situation. However, the main reason may lie in the recent dramatic development of computing capabilities, particularly in the form of microprocessors, which have enabled us to perform symbolic manipulations in a way hitherto impossible. As a result, the complexity of the systems modeled and the corresponding mathematical models have radically increased, a complexity that has stimulated numerous contributions to support mathematical modeling activities. The main reason of many important applications of mathematical modeling in various scientific disciplines is the dramatic evolution of computers over the last 30 years. Today mathematical modelers can perform symbolic manipulations in such ways that these modelers could not imagine three decades ago. The literature of modern mathematical modeling includes plethora of excellent research publications which illustrate the very strong possibilities of the combination of mathematical methodology and computational power. Result of this combination is the increase of complexity of modeled real systems and the corresponding mathematical models. These increases are challenges for the existing processes of mathematical modeling and create opportunities for producing new research publications supporting the mathematical modeling processes.

Many mathematical models are deterministic in the sense that they always give the same output when a specific input is applied. A chance mechanism is included in stochastic models. It has been observed that as a system, be it social, economic or physical, gets more complex, it becomes more difficult to be described by deterministic mathematical models. In fact, probability theory and stochastic models were developed in response to the failure of deterministic models to deal with complicated systems. It is generally recognized that stochastic models can be applied to practical situations in many ways. For the results to be applicable stochastic models should be used in a region where they are valid.

It is generally accepted that stochastic models can be applied to practical situations in many ways. However, the successful implementation of a stochastic model in a practical situation depends on the suitability of this model to describe the particular practical situation. The ascertainment that a stochastic model provides a suitable description of a real situation is an extremely difficult work, which requires that the modeler must be very careful. Practical experience has shown that it is an inevitable requirement the verification of the suitability of a stochastic model has to be repeated many times by various experts in stochastic modeling matters. If the verification of the reliability of a stochastic model is not possible then the application of this model to the description and analysis of real situations should be avoided.

The quantitative character of risk severity, risk frequency, risk occurrence time, risk duration, and risk occurrence space and other fundamental components of the complex concept of risk permit the use of mathematical models in risk management problems. Moreover, since the quantitative components of risk are random variables then these models are stochastic. The last 30 years the developments in the discipline of risk management are impressive. These developments are largely due to the use of stochastic models in describing, analyzing, and implementing the safety operations of a firm, namely risk identification, risk measurement, and risk treatment.

A thorough examination of stochastic modeling in the development of risk management as an important function of a firm reveals the very significant role of basic and mixed probability distributions in the implementation of risk measurement and risk treatment. Since the evaluation of the probability distribution of a stochastic model is a very difficult and sometimes impossible procedure, then it is appropriate the interpretation of a given basic or mixed probability distribution as the probability distribution of a stochastic model which is used to describe and analyze problems or situations in the area of risk management.

The present section of this work presents the theoretical and practical possibilities of stochastic modeling and stochastic models for shaping risk management as a particular important discipline of general management. More precisely, the present section makes quite clear that stochastic models are very strong analytical tools for implementing the risk management operations of an organization. Moreover, this section interprets the significant role of basic and mixed probability distributions in stochastic modeling and stochastic models.

Risk managers and experts in formulating and applying stochastic models are very often involved in making forecasts for the results of random procedures and activities. This means that understanding of the uncertainty constitutes the most crucial factor for the realization of principles, operations, and strategic objectives of risk management. Moreover, it is very important risk managers and other risk experts to recognize and accurately transmit the existing uncertainty. During the last four decades, the bibliography and practice of risk management recognize probability theory as the most suitable discipline for the mathematical representation of uncertainty. This implies that stochastic modeling and stochastic models have a significant role in the very wide area of practical applications of risk management.

Risk managers and other risk experts obtain valuable information concerning selection and implementation of safety operations by making use of stochastic models. The bibliography of risk management frequently makes use of the term "risk management operations" instead of the term "safety operations".

The existence of stochastic models available for describing real problems in the area of risk management enables the evaluation of alternative courses of action when designing a proactive risk management program, which is a key factor for the implementation of the strategic goals of a firm. The clarification of the role of time in formulating a stochastic model for risk management is very important. Many problems in the area of risk management can be described and solved by static stochastic models, or equivalently by stochastic models which are free of the concept of time. Examples of static stochastic models for risk management are the tree diagrams of probabilistic analysis of safety of various systems. The elimination of the concept of time from these tree diagrams is made for more accurate modeling of the structural elements of the system. Also many problems in the area of risk management can be described, analyzed, and solved by making use of dynamic stochastic models, or equivalently by stochastic models that include the concept of time. The practical usefulness of dynamic stochastic models of risk management is greater than the practical usefulness of static stochastic models of risk management. However the formulation, investigation, and implementation of dynamic stochastic models of risk management are more difficult than the formulation, investigation, and implementation of static stochastic models of risk management. Static and dynamic stochastic models of risk management are based on many assumptions that must be checked very carefully in order to avoid false interpretations of these stochastic models. Moreover, every stochastic model of risk management always involves an unknown relationship between its random variables which relationship has the potential to generate a sequence of random events with catastrophic consequences.

The stochastic consideration of risk management is based on the general problem of decision making under conditions of uncertainty. Traditionally this problem has been addressed by the theory of expected utility. This theory is of course very limited for the description and treatment of real situations, but the theory of expected utility helps internal consideration of decision making under conditions of uncertainty by focusing the attention of researchers on the validity of the axioms of the theory. Decision making under conditions of uncertainty makes use of stochastic models for predicting the results of specific managerial actions.

The international scientific community of risk management has indicated a very significant interest in stochastic models. These models are used in many areas of risk management since risk managers, and other risk experts, work in an environment which is extremely complex and uncertain. The fact that stochastic models enable the users of these models to isolate and study the various cognitive procedures involved implies that risk managers can significantly improve the decision making processes to treat risks that threaten the efficient function and integrity of various firms. The quantitative information for the risks threatening various firms is absolutely necessary if risk managers are to take effective decisions for controlling and financing these risks. The stochastic models of risk management are very important for the analysis of such quantitative information. The mystery surrounding the application of stochastic models to problems occurring at the area of risk management prevents risk managers and other risk experts to successfully implement these models. Risk managers argue that processes of making and implementing effective decisions in various areas of risk management should be based on stochastic models which are the results of hard and lengthy efforts. It is generally accepted that very good managerial skills are necessary for the successful implementation of risk management objectives. However, without the use of effective risk management stochastic models then the successful implementation of the objectives of risk management is impossible.

During the last 30 years risk management became global, systemic, and proactive. The fundamental risk management operations, namely risk identification, risk measurement, and risk treatment have shown significant improvement due to the extensive use of stochastic models. A very interesting research area of stochastic modeling is the formulation, investigation, and application of stochastic models for the description, analysis, and control of risk management operations incorporating the concepts of risk severity, risk frequency, and risk duration. The main purpose of the present work is the formulation and investigation of stochastic models incorporating the risk severity, risk frequency, and risk duration. The particular practical significance of these quantitative components of the concept of risk is the main reason for the consideration of the stochastic models of this work.

The recognition of probability theory as a fundamental structural factor of risk management constitutes a very important reason for undertaking research activities in the area of probability distributions arising in stochastic modeling of safety operations. A description of the role of stochastic models in risk management problems makes quite clear the importance of the applications of probability distributions in the fundamental safety operations. These applications make use of very strong results of probability theory and provide risk managers with very useful information for risk management decision making.

Empirical investigations concentrating on the applications of risk management in modern large organizations make quite clear that risk managers and other risk experts substantially support the formulation, investigation, and use of stochastic models for the treatment of new risks threatening these organizations. More precisely, for the treatment of new risks related to space exploration, engineering, and use of computers.

Chapter 2
Stochastic Models of Risk Management Concepts

Abstract The formulation and investigation of stochastic models for the fundamental quantitative concepts of risk management constitute the purpose of this chapter. The concepts of the main quantitative components of risk and the concepts of the main quantitative components of risk control and risk financing operations constitute the fundamental quantitative concepts of risk management. This chapter consists of two parts. The first part concentrates on the formulation and investigation of stochastic models for risk severity, risk duration, risk frequency, and total risk severity which are the main quantitative components of risk. The second part concentrates on the formulation and investigation of stochastic models for the time required for treating a risk occurrence, the time of the first occurrence of a major risk, the minimum time of a random number of risk occurrences, the number of ongoing risk occurrences, the multiplicative risk severity, and the riskiness which are the main quantitative components of risk control and risk financing operations.

2.1 Introduction

The purpose of the present chapter is the formulation and investigation of stochastic models for the fundamental quantitative concepts of the discipline of risk management. These concepts include the concepts of the main quantitative components of risk and the concepts of the main quantitative components of risk control and risk financing operations. Risk severity, risk duration, risk frequency, and total risk severity are the main quantitative components of risk. The concepts of the main quantitative components of risk and the stochastic models of these concepts constitute the first part of the present chapter. The time required for treating a risk occurrence, the time of the first occurrence of a major risk, the minimum time of a random number of risk occurrences, the number of ongoing risk occurrences, the multiplicative risk severity, and the riskiness are the main quantitative components of risk control and risk financing operations. The concepts of the main quantitative components of risk control and risk financing operations and the stochastic models of these concepts constitute the second part of the present

C. Artikis and P. Artikis, *Probability Distributions in Risk Management Operations*,
Intelligent Systems Reference Library 83, DOI 10.1007/978-3-319-14256-2_2

chapter. The stochastic models of the fundamental quantitative concepts of risk management are generally recognized as very strong analytical tools for investigating problems and making decisions in various areas of this discipline. These models provide risk experts with valuable information for the implementation of risk management principles. The implementation of risk management principles is realized with the contribution of the stochastic models of the main quantitative components of risk as structural elements in stochastic modeling activities of the main quantitative components of risk control and risk financing operations. In conclusion, the importance of the stochastic models of the main quantitative components of risk substantially supports the importance of the stochastic models of the main quantitative components of risk control and risk financing operations.

The investigation of the stochastic models of the fundamental concepts of risk management uses the results of the theory of mixed probability distributions. In particular, the very strong results of the theory of characteristic functions corresponding to mixed probability distributions are very useful for investigating properties of such stochastic models. The establishment of unimodality, infinite divisibility, selfdecomposability, and other properties for the probability distributions of the stochastic models describing the fundamental concepts of risk management is facilitated by the use of the corresponding characteristic functions.

2.2 Model of Risk Severity

The economic loss due to the occurrence of a risk is defined as risk severity. The suitable stochastic model for the description and analysis of risk severity is a continuous random variable X with values in the interval $(0, \infty)$.

The distribution function $F_X(x)$, the probability density function $f_X(x)$, and the characteristic function $\varphi_X(u)$, $u \in \mathbf{R}$ of the random variable X are the concepts of probability theory which constitute the basic analytical tools for investigating the behaviour of risk severity. The establishment of infinite divisibility, selfdecomposability, unimodality and other properties of the probability distribution of risk severity substantially supports the activities of developing, investigating, and applying of various risk control operations. These operations are the structural elements of proactive risk management which is generally recognized as the modern perspective of this discipline.

2.3 Model of Risk Duration

The length of the time interval in which the cause of a risk creates economic loss is defined as risk duration. The suitable stochastic model for the description and analysis of risk duration is a continuous random variable S with values in the interval $(0, \infty)$.

The distribution function $F_S(s)$, the probability density function $f_S(s)$, and the characteristic function $\varphi_S(u)$ of the random variable S are the concepts of probability theory which constitutes the basic analytical tools for investigating the behavior of risk duration. The establishment of theoretical properties for the probability distribution of risk duration substantially supports the activities of developing, investigating, and applying of various risk control operations. Risk duration can be used in the formulation of stochastic multiplicative models for the description and analysis of risk severity.

2.4 Model of Risk Frequency

The number of the occurrences of a risk in a given time interval is defined as risk frequency. The suitable stochastic model for the description and analysis of risk frequency is a discrete random variable N with values in the set $\mathbf{N}_0 = \{0, 1, 2, \ldots\}$.

The probability function $P(N = n) = p_n, n = 0, 1, 2, \ldots$ and the probability generating function $P_N(z), |z| \leq 1$ of the random variable N are the concepts of probability theory which constitute the basic analytical tools for investigating the behavior of risk frequency. The establishment of theoretical properties for the probability distribution of risk frequency supports the activities of developing, investigating, and applying of various risk control operations. Risk frequency is particularly useful in formulating stochastic models for the description and analysis of total risk severity. These models are structural elements of risk control and risk financing operations.

The consideration of risk frequency in a time interval of the form $[0, t]$ is of significant practical interest. In this interval, risk frequency is represented by the random variable $N(t)$ with probability generating function

$$P_{N(t)}(z, t). \tag{2.4.1}$$

In this case $\{N(t), t \geq 0\}$ is a counting stochastic process.

The consideration of risk frequency in a time interval of the form $[0, T]$, where T is a continuous random variable with probability density function

$$f_T(t), \tag{2.4.2}$$

is particularly useful in a wide variety of practical disciplines. The risk frequency in the time interval $[0, T]$ is represented by the random variable K.

The probability generating function of this random variable has the form

$$P_K(z) = E\big(E(z^K|T)\big). \tag{2.4.3}$$

From (2.4.2) and (2.4.3) it follows that

$$P_K(z) = \int_0^\infty E(z^K | T = t) f_T(t) dt. \tag{2.4.4}$$

Since the random variable $K|T = t$ is equally distributed with the random variable $N(t)$ then we get that

$$E(z^K | T = t) = E\left(z^{N(t)}\right). \tag{2.4.5}$$

From (2.4.1) and (2.4.5) we get that

$$E(z^K | T = t) = P_{N(t)}(z, t). \tag{2.4.6}$$

If we use (2.4.6) in (2.4.4) then the probability generating function of the random variable K has the form

$$P_K(z) = \int_0^\infty P_{N(t)}(z, t) f_T(t) dt. \tag{2.4.7}$$

If the counting stochastic process $\{N(t), t \geq 0\}$ is a homogeneous Poisson process with probability generating function $P_{N(t)}(z, t) = e^{\lambda t(z-1)}, \lambda > 0$, then (2.4.7) has the form

$$P_K(z) = \int_0^\infty e^{\lambda t(z-1)} f_T(t) dt. \tag{2.4.8}$$

Since the homogeneous Poisson process is the most important counting process, from a theoretical and a practical point of view, and the discrete distributions with probability generating functions of the form (2.4.8) are strong tools of probability theory then the evaluation of special cases of (2.4.8) is very interesting.

If the random variable T follows the uniform distribution with probability density function

$$f_T(t) = 1, \quad 0 < t < 1, \tag{2.4.9}$$

then from (2.4.8) and (2.4.9) it follows that the discrete random variable K, which denotes the risk frequency in the time interval $[0, T]$, has probability generating function $P_K(z) = \int_0^1 e^{\lambda t(z-1)} dt$ or equivalently

$$P_K(z) = \frac{1 - e^{\lambda(z-1)}}{\lambda(1-z)}. \tag{2.4.10}$$

The probability generating function (2.4.10) belongs to the renewal distribution which corresponds to the Poisson distribution.

If the random variable T follows the exponential distribution with probability density function

$$f_T(t) = \mu e^{-\mu t},\ t > 0,\ \mu > 0, \tag{2.4.11}$$

then from (2.4.8) and (2.4.11) it follows that the random variable K, which denotes the risk frequency in the time interval $[0, T]$, has probability generating function

$$P_K(z) = \int_0^\infty e^{\lambda t(z-1)} \mu e^{-\mu t} dt$$

or equivalently

$$P_K(z) = \frac{\mu}{\mu + \lambda(1-z)}. \tag{2.4.12}$$

If we set $p = \frac{\mu}{\mu+\lambda}$ and $q = \frac{\lambda}{\lambda+\mu}$, then the probability generating function (2.4.12) has the form

$$P_K(z) = \frac{p}{1 - qz}. \tag{2.4.13}$$

The probability generating function (2.4.13) belongs to the geometric type I distribution.

We suppose that the Laplace transform of the random variable T is

$$\omega(\rho) = \int_0^\infty e^{-\rho t} f_T(t) dt,\ \rho \geq 0. \tag{2.4.14}$$

Since the probability generating function (2.4.8) has the form

$$P_K(z) = \int_0^\infty e^{-\lambda t(1-z)} f_T(t) dt, \tag{2.4.15}$$

then from (2.4.14) and (2.4.15) it follows that

$$P_K(z) = \omega(\lambda(1-z)). \tag{2.4.16}$$

A particular case of (2.4.16) is the following. We suppose that the distribution of the random variable T belongs to the class of continuous stable distributions with Laplace transform

$$\omega(\rho) = e^{-\rho^{\gamma}},\ 0 < \gamma \leq 1. \tag{2.4.17}$$

From (2.4.16) and (2.4.17) it follows that the random variable K, which denotes the frequency of risk in the time interval $[0,T]$, has probability generating function

$$P_K(z) = e^{-\lambda^{\gamma}(1-z)^{\gamma}}. \tag{2.4.18}$$

The probability generating function (2.4.18) belongs in a distribution which is a member of the class of discrete stable distributions. The class of discrete stable distributions is very important, in theory and practice, for five reasons. The first reason is the unimodality of the discrete stable distributions. The infinite divisibility of the discrete stable distributions is the second reason. The fact that the class of discrete stable distributions includes distributions of significant theoretical and practical interest constitutes the third reason. An example of such distribution is the Poisson distribution with probability generating function $P_K(z) = e^{\lambda(z-1)}$ being of the form (2.4.18) with $\gamma = 1$.

The fourth reason is the representation of a discrete random variable L, with values in the set $\mathbf{N}_0 = \{0,1,2,\ldots\}$ and distribution belonging to the class of discrete stable distributions, as a Poisson random sum of random variables following the Sibuya distribution. The representation is implemented in the following way. Since the distribution of the random variable L belongs to the class of discrete stable distributions then the probability generating function of the random variable L has the form

$$P_L(z) = e^{-c(1-z)^{\gamma}},\ c > 0,\ 0 < \gamma \leq 1. \tag{2.4.19}$$

The probability generating function (2.4.18) is of the form (2.4.19) with $c = \lambda^{\gamma}$.

We consider the discrete random variable E which follows the Poisson distribution with probability generating function $P_E(z) = e^{c(z-1)}$ and the sequence of independent random variables $\{X_{\varepsilon}, \varepsilon = 1,2,\ldots\}$.

The random variable E is independent of the sequence $\{X_{\varepsilon}, \varepsilon = 1,2,\ldots\}$ of random variables distributed as the random variable X which follows the Sibuya distribution with probability generating function $P_X(z) = 1-(1-z)^{\gamma}$.

Since the probability generating function $P_L(z) = e^{-c(1-z)^{\gamma}}$ has the form $P_L(z) = e^{c[1-(1-z)^{\gamma}-1]}$ then the random variable L has the form of a Poisson random sum of random variables following the Sibuya distribution, or equivalently the random variable L has the form $L = X_1 + X_2 + \cdots + X_E$.

The fifth reason is the construction of important mixed distributions with the use of the class of discrete stable distributions.

2.5 Total Risk Severity

The discrete random variable N with values in the set $\mathbf{N}_0 = \{0, 1, 2, \ldots\}$ and probability generating function $P_N(z)$ is independent of the sequence of continuous, positive, independent, and identically distributed random variables $\{X_n, n = 1, 2, \ldots\}$.

The random variables of the above sequence are equally distributed with the random variable X which has characteristic function $\varphi_X(u)$.

If the random variable N represents the frequency of a risk and the random variable X_n represents the severity of the nth risk occurrence then the random sum $L = X_1 + X_2 + \cdots + X_N$ represents the total risk severity. The study of the total risk severity is based on the characteristic function $\varphi_L(u)$ since it is not possible the evaluation of the distribution function $F_L(\ell)$ and the evaluation of the probability density function $f_L(\ell)$.

The characteristic function $\varphi_L(u)$ is evaluated in the following way. We have that $\varphi_L(u) = E(e^{iuL})$ or equivalently

$$\varphi_L(u) = E\big(E(e^{iuL}|N)\big). \tag{2.5.1}$$

From (2.5.1) it follows that

$$\varphi_L(u) = \sum_{n=0}^{\infty} E\big(e^{iuL}|N = n\big)\ P(N = n)$$

or equivalently

$$\varphi_L(u) = \sum_{n=0}^{\infty} E\big(e^{iuX_1 + \cdots + iuX_n}|N = n\big)P(N = n). \tag{2.5.2}$$

From (2.5.2) it follows that

$$\varphi_L(u) = \sum_{n=0}^{\infty} E\big(e^{iuX_1} \cdots e^{iuX_n}|N = n\big)\ P(N = n). \tag{2.5.3}$$

The independence of the random variable N from the sequence of random variables $\{X_n, n = 1, 2, \ldots\}$ implies the independence of the random variables $N, X_1, \ldots, X_n$.

Hence the random variables $N, e^{iuX_1}, \ldots, e^{iuX_n}$ are independent and the random variables $e^{iuX_1}, \ldots, e^{iuX_n}$ are also independent. It is easily seen that (2.5.3) has the form

$$\varphi_L(u) = \sum_{n=0}^{\infty} E(e^{iuX_1}) \ldots E(e^{iuX_n}) P(N = n). \tag{2.5.4}$$

Since

$$\varphi_X(u) = E(e^{iuX_n}),\ n = 1, 2, \ldots$$

then (2.5.4) has the form

$$\varphi_L(u) = \sum_{n=0}^{\infty} \varphi_X^n(u) P(N = n). \tag{2.5.5}$$

From (2.5.5) it follows that the characteristic function of the total risk severity has the form

$$\phi_L(u) = P_N(\phi_X(u)). \tag{2.5.6}$$

The characteristic function (2.5.6), the theorem of inversion of characteristic functions, and the Fast Fourier Transform algorithm make possible the study of the probabilistic behavior of the total risk severity. This behavior is important for making decisions in the areas of risk control and risk financing operations.

The mean value of the total risk severity $E(L)$ can be used in risk classification operations. The mean value of the total risk severity can be evaluated in the following way. From (2.5.6) we get that

$$\varphi_L'(u) = \varphi_X'(u) P_N'(\varphi_X(u)). \tag{2.5.7}$$

Hence (2.5.7) implies that

$$\varphi_L'(0) = \varphi_X'(0) P_N'(\varphi_X(0)). \tag{2.5.8}$$

From (2.5.8) it follows that

$$E(L) = E(X) E(N). \tag{2.5.9}$$

A special case of the total risk severity, with significant practical interest, is the following.

We suppose that the frequency of a risk follows the Bernoulli distribution with probability function $P(N = n) = p^n q^{1-n}, n = 0, 1$.

The probability generating function of the random variable N is $P_N(z) = q + pz$.

The mean value of the random variable N is $E(N) = p$.

We suppose that the severity of a risk is a continuous and positive random variable X with characteristic function $\varphi_X(u)$ and mean value $E(X)$.

The total severity of risk is the random variable

$$L = \begin{cases} 0, & N = 0 \\ X, & N = 1. \end{cases}$$

The characteristic function of the total risk severity L is $\varphi_L(u) = q + p\varphi_X(u)$.

From (2.5.9) we get that the mean value of the total severity L is $E(L) = pE(X)$.

An extension of the random sum $L = X_1 + X_2 + \cdots + X_N$ as a model of total risk severity is the following. Let N be a discrete random variable with values in the set $\mathbf{N}_0 = \{0, 1, 2, \ldots\}$ and probability generating function $P_N(z)$.

Let $\{V_n, n = 1, 2, \ldots\}$ be a sequence of discrete and independent random variables, distributed as the random variable V with values in the set $\mathbf{N}_0 = \{0, 1, 2, \ldots\}$ and probability generating function $P_V(z)$.

We suppose that the random variable N is independent of the sequence of random variables $\{V_n, n = 1, 2, \ldots\}$ and we set $K = V_1 + V_2 + \cdots + V_N$.

Let $\{X_\kappa, \kappa = 1, 2, \ldots\}$ be a sequence of continuous, positive and independent random variables, distributed as the random variable X with characteristic function $\varphi_X(u)$, and we set $L = X_1 + X_2 + \cdots + X_K$.

An interpretation of the random variable $L = X_1 + X_2 + \cdots + X_K$ in the area of stochastic models of total risk frequency is the following.

We suppose that the random variable N denotes the frequency of a risk and the random variable V_n denotes the number of different damages due to the nth risk occurrence. The random variable $K = V_1 + V_2 + \cdots + V_N$ denotes the number of different damages due to the N risk occurrences. The random variable X_κ denotes the size of the κth damage. Hence the random variable $L = X_1 + X_2 + \cdots + X_K$ denotes the total severity of risk.

The following result establishes sufficient conditions for the evaluation of the characteristic function $\varphi_L(u)$ of the random variable $L = X_1 + X_2 + \cdots + X_K$.

Theorem 2.5.1 *Let N be a discrete random variable with values in the set $\mathbf{N}_0 = \{0, 1, 2, \ldots\}$ and probability generating function $P_N(z)$.*

Let $\{V_n, n = 1, 2, \ldots\}$ be a sequence of discrete and independent random variables, distributed as the random variable V with values in the set $\mathbf{N}_0 = \{0, 1, 2, \ldots\}$, and probability generating function $P_V(z)$.

We set

$$K = V_1 + V_2 + \cdots + V_N.$$

Let $\{X_\kappa, \kappa = 1, 2, \ldots\}$ be a sequence of continuous, positive, and independent random variables distributed as the random variable X with characteristic function $\varphi_X(u)$.

We set

$$L = X_1 + X_2 + \cdots + X_K.$$

If N, $\{V_n, n = 1, 2, \ldots\}$ and $\{X_\kappa, \kappa = 1, 2, \ldots\}$ are independent then the characteristic function of the random variable $L = X_1 + X_2 + \cdots + X_K$ is $\varphi_L(u) = P_N(P_V(\varphi_X(u)))$.

Proof We have $\varphi_L(u) = E(e^{iuL})$ or equivalently

$$\varphi_L(u) = E\big(E(e^{iuL}|K)\big). \tag{2.5.10}$$

From (2.5.10) we get that

$$\varphi_L(u) = \sum_{\kappa=0}^{\infty} E(e^{iuL}|K = \kappa)P(K = \kappa)$$

or equivalently we get that

$$\varphi_L(u) = \sum_{\kappa=0}^{\infty} E\left(e^{iu(X_1+\cdots+X_K)}|K = \kappa\right) P(K = \kappa). \tag{2.5.11}$$

From (2.5.11) it follows that

$$\varphi_L(u) = \sum_{\kappa=0}^{\infty} E\left(e^{iu(X_1+\cdots+X_\kappa)}|K = \kappa\right) P(K = \kappa). \tag{2.5.12}$$

Since $K = V_1 + V_2 + \cdots + V_N$ then (2.5.12) has the form

$$\varphi_L(u) = \sum_{\kappa=0}^{\infty} E\left(e^{iu(X_1+\cdots+X_\kappa)}|V_1 + V_2 + \cdots + V_N = \kappa\right)P(V_1 + V_2 + \cdots + V_N = \kappa). \tag{2.5.13}$$

We shall prove that the random variables $\Lambda = X_1 + X_2 + \cdots + X_\kappa$ and $K = V_1 + V_2 + \cdots + V_N$ are independent. If $\varphi_\Lambda(u)$ is the characteristic function of the random variable $\Lambda = X_1 + X_2 + \cdots + X_\kappa$ then we get that $\varphi_\Lambda(u) = E(e^{iu\Lambda})$ or equivalently we get that

$$\varphi_\Lambda(u) = E\left(e^{iu(X_1+\cdots+X_\kappa)}\right). \tag{2.5.14}$$

From (2.5.14) it follows that

$$\varphi_{\Lambda}(u) = E\left(e^{iuX_1+\cdots+iuX_\kappa}\right). \quad (2.5.15)$$

Since the random variables of the sequence $\{X_\kappa, \kappa = 1, 2, \ldots\}$ are independent then the random variables $X_1, \ldots, X_\kappa$ are independent. The independence of the above random variables implies the independence of the random variables $e^{iuX_1}, \ldots, e^{iuX_\kappa}$.

Hence (2.5.15) has the form

$$\varphi_{\Lambda}(u) = E\left(e^{iuX_1}\right)\ldots E\left(e^{iuX_\kappa}\right). \quad (2.5.16)$$

Since the random variables of the sequence $\{X_\kappa, \kappa = 1, 2, \ldots\}$ are equally distributed with the random variable X having characteristic function $\varphi_X(u)$ then (2.5.16) has the form $\varphi_{\Lambda}(u) = \varphi_X^{\kappa}(u)$.

Let $\varphi_K(\xi)$ be the characteristic function of the random variable $K = V_1 + V_2 + \cdots + V_N$.

The independence of $\{V_n, n = 1, 2, \ldots\}$, N and $\{X_\kappa, \kappa = 1, 2, \ldots\}$ implies the independence of $\{V_n, n = 1, 2, \ldots\}$ and N.

Since the random variables of the sequence $\{V_n, n = 1, 2, \ldots\}$ are independent and equally distributed with the random variable V and since the random variable N has probability generating function $P_N(z)$ then from (2.5.6) it follows that the characteristic function of the random variable $K = V_1 + V_2 + \cdots + V_N$ is $\varphi_K(\xi) = P_N(\varphi_V(\xi))$ where $\varphi_V(\xi)$ is the characteristic function of the random variable V.

The proof of the independence of the random variables $\Lambda = X_1 + X_2 + \cdots + X_\kappa$ and $K = V_1 + V_2 + \cdots + V_N$ requires the establishment of the relationship $\varphi_{\Lambda,K}(u, \xi) = \varphi_{\Lambda}(u)\varphi_K(\xi)$ where $\varphi_{\Lambda,K}(u, \xi)$ is the characteristic function of the vector (Λ, K).

We have $\varphi_{\Lambda,K}(u, \xi) = E\left(e^{iu\Lambda+i\xi K}\right)$ or equivalently we have $\varphi_{\Lambda,K}(u, \xi) = E\left(E\left(e^{iu\Lambda+i\xi K}|N\right)\right)$.

Hence

$$\varphi_{\Lambda,K}(u, \xi) = \sum_{n=0}^{\infty} E\left(e^{iu(X_1+X_2+\cdots+X_\kappa)+i\xi K}|N = n\right)P(N = n). \quad (2.5.17)$$

From (2.5.17) it follows that

$$\varphi_{\Lambda,K}(u, \xi) = \sum_{n=0}^{\infty} E\left(e^{iu(X_1+X_2+\cdots+X_\kappa)+i\xi(V_1+V_2+\cdots+V_N)}|N = n\right)P(N = n)$$

or equivalently it follows that

$$\varphi_{\Lambda,K}(u,\xi) = \sum_{n=0}^{\infty} E\Big(e^{iu(X_1+X_2+\cdots+X_\kappa)+i\xi(V_1+V_2+\cdots+V_n)}|N=n\Big)P(N=n). \quad (2.5.18)$$

From (2.5.18) it follows that

$$\varphi_{\Lambda,K}(u,\xi) = \sum_{n=0}^{\infty} E(e^{iuX_1}\ldots e^{iuX_\kappa}e^{i\xi V_1}\ldots e^{i\xi V_n}|N=n)P(N=n). \quad (2.5.19)$$

The independence of $\{V_n, n=1,2,\ldots\}$, N and $\{X_\kappa, \kappa=1,2,\ldots\}$ implies the independence of the random variables $V_1, \ldots, V_n, X_1, \ldots, X_\kappa, N$.

The independence of the above random variables implies the independence of the random variables $e^{i\xi V_1}, \ldots, e^{i\xi V_n}, e^{iuX_1}, \ldots, e^{iuX_\kappa}, N$.

Hence (2.5.19) has the form

$$\varphi_{\Lambda,K}(u,\xi) = \sum_{n=0}^{\infty} E(e^{iuX_1}\ldots e^{iuX_\kappa}e^{i\xi V_1}\ldots e^{i\xi V_n})P(N=n). \quad (2.5.20)$$

The independence of the random variables $e^{iuX_1}, \ldots, e^{iuX_\kappa}, e^{i\xi V_1}, \ldots, e^{i\xi V_n}, N$ implies the independence of the random variables $e^{iuX_1}, \ldots, e^{iuX_\kappa}, e^{i\xi V_1}, \ldots, e^{i\xi V_n}$.

Hence (2.5.20) has the form

$$\varphi_\Lambda(u) = \sum_{n=0}^{\infty} E(e^{iuX_1})\ldots E(e^{iuX_\kappa})E(e^{i\xi V_1})\ldots E(e^{i\xi V_n})P(N=n). \quad (2.5.21)$$

Since the random variables of the sequence $\{X_\kappa, \kappa=1,2,\ldots\}$ are equally distributed with the random variable X having characteristic function $\varphi_X(u)$ and the random variables of the sequence $\{V_n, n=1,2,\ldots\}$ are equally distributed with the random variable V with characteristic function $\varphi_V(u)$ then (2.5.21) has the form

$$\varphi_{\Lambda,K}(u,\xi) = \varphi_X^\kappa(u)\sum_{n=0}^{\infty}\varphi_V^n(\xi)P(N=n)$$

or equivalently the form

$$\varphi_{\Lambda,K}(u,\xi) = \varphi_X^\kappa(u)P_N(\varphi_V(\xi)). \quad (2.5.22)$$

Hence (2.5.22) has the form $\varphi_{\Lambda,K}(u,\xi) = \varphi_\Lambda(u)\varphi_K(\xi)$, which means that the random variables $\Lambda = X_1 + X_2 + \cdots + X_\kappa$ and $K = V_1 + V_2 + \cdots + V_N$ are independent. The independence of the above random variables implies that (2.5.13) has the form

$$\varphi_L(u) = \sum_{\kappa=0}^{\infty} E\left(e^{iu(X_1+\cdots+X_\kappa)}\right) P(V_1 + V_2 + \cdots + V_N = \kappa)$$

or equivalently the form

$$\varphi_L(u) = \sum_{\kappa=0}^{\infty} \varphi_X^\kappa(u) P(V_1 + V_2 + \cdots + V_N = \kappa). \tag{2.5.23}$$

Since the random variables of the sequence $\{V_n, n = 1, 2, \ldots\}$ are equally distributed with the random variable V having probability generating function $P_V(z)$ and $P_N(z)$ is the probability generating function of the random variable N.

Then from the form (2.5.6), for probability generating functions, it follows that the probability generating function of the random sum $K = V_1 + V_2 + \cdots + V_N$ is $P_K(z) = P_N(P_V(z))$.

Since

$$P_K(z) = \sum_{\kappa=0}^{\infty} z^\kappa P(K = \kappa)$$

or equivalently

$$P_N(P_V(z)) = \sum_{\kappa=0}^{\infty} z^\kappa P(V_1 + V_2 + \cdots + V_N = \kappa)$$

then (2.5.23) implies that the characteristic function of the random sum $L = X_1 + X_2 + \cdots + X_K$ with $K = V_1 + V_2 + \cdots + V_N$ is

$$\varphi_L(u) = P_N(P_V(\varphi_X(u))). \tag{2.5.24}$$

The consideration of special cases of the probability generating function (2.5.24) when the random variable N follows the Poisson distribution is very important because the Poisson distribution is the most usual distribution of risk frequency.

We suppose that the random variable N follows the Poisson distribution with probability generating function

$$P_N(z) = e^{\lambda(z-1)} \tag{2.5.25}$$

and the random variable V follows the Poisson distribution with probability generating

$$P_V(z) = e^{\theta(z-1)} \tag{2.5.26}$$

then from (2.5.24), (2.5.25) and (2.5.26) it follows that the probability generating function $P_K(z) = P_N(P_V(z))$ of the random sum $K = V_1 + V_2 + \cdots + V_N$ has the form

$$P_K(z) = \exp\left[\lambda\left(e^{\theta(z-1)} - 1\right)\right]. \tag{2.5.27}$$

Hence the random sum $K = V_1 + V_2 + \cdots + V_N$ follows the Neyman type A distribution. We suppose that the random variable X follows the exponential distribution with characteristic function

$$\varphi_X(u) = \frac{\mu}{\mu - iu}. \tag{2.5.28}$$

From (2.5.24), (2.5.27) and (2.5.28) it follows that the characteristic function of the random sum $L = X_1 + X_2 + \cdots + X_K$ has the form

$$\varphi_L(u) = \exp\left\{\lambda\left[e^{\theta\left(\frac{\mu}{\mu-iu}-1\right)} - 1\right]\right\}.$$

If the random variable N follows the Poisson distribution with probability generating function

$$P_N(z) = e^{\lambda(z-1)} \tag{2.5.29}$$

and the random variable V follows the binomial distribution with probability generating function

$$P_V(z) = (pz + q)^m \tag{2.5.30}$$

then from (2.5.24), (2.5.29) and (2.5.30) it follows that the probability generating function $P_K(z) = P_N(P_V(z))$ of the random sum $K = V_1 + V_2 + \cdots + V_N$ has the form

$$P_K(z) = \exp\{\lambda[(pz + q)^m - 1]\}. \tag{2.5.31}$$

Hence the random sum $K = V_1 + V_2 + \cdots + V_N$ follows the Poisson – binomial distribution. We suppose that the random variable X follows the uniform distribution with characteristic function

$$\varphi_X(u) = \frac{e^{iu} - 1}{iu}. \tag{2.5.32}$$

From (2.5.24), (2.5.31) and (2.5.32) it follows that the characteristic function of the random sum $L = X_1 + X_2 + \cdots + X_K$ has the form

$$\varphi_L(u) = \exp\left\{ \lambda \left[\left(p \frac{e^{iu} - 1}{iu} + q \right)^m - 1 \right] \right\}.$$

If the random variable N follows the Poisson distribution with probability generating function

$$P_N(z) = e^{\lambda(z-1)} \tag{2.5.33}$$

and the random variable V follows the geometric type II distribution with probability generating function

$$P_V(z) = \frac{pz}{1 - qz} \tag{2.5.34}$$

then from (2.5.33) and (2.5.34) it follows that the probability generating function $P_K(z) = P_N(P_V(z))$ of the random sum $K = V_1 + V_2 + \cdots + V_N$ has the form

$$P_K(z) = \exp\left\{ \lambda \left[\frac{pz}{1 - qz} - 1 \right] \right\}. \tag{2.5.35}$$

Hence the random sum $K = V_1 + V_2 + \cdots + V_N$ follows the Polya-Aeppli distribution. We suppose that the random variable X follows the exponential distribution with characteristic function

$$\varphi_X(u) = \frac{\mu}{\mu - iu}. \tag{2.5.36}$$

From (2.5.24), (2.5.35) and (2.5.36) it follows that the characteristic function of the random sum $L = X_1 + X_2 + \cdots + X_K$ has the form

$$\varphi_L(u) = \exp\left\{ \lambda \left[\frac{p\mu}{p\mu - iu} \right] - 1 \right\}.$$

We suppose that the random variable N follows the Poisson distribution with probability generating function

$$P_N(z) = e^{\lambda(z-1)} \tag{2.5.37}$$

and the random variable V has the form $V = \Pi + 1$, where Π is a random variable following the Poisson distribution with probability generating function $P_\Pi(z) = e^{\theta(z-1)}$.

The probability generating function of the random variable V is

$$P_V(z) = ze^{\theta(z-1)}. \tag{2.5.38}$$

From (2.5.24), (2.5.37) and (2.5.38) it follows that the probability generating function $P_K(z) = P_N(P_V(z))$ of the random sum $K = V_1 + V_2 + \cdots + V_N$ has the form

$$P_K(z) = \exp\left\{\lambda\left[ze^{\theta(z-1)} - 1\right]\right\}. \tag{2.5.39}$$

Hence the random sum $K = V_1 + V_2 + \cdots + V_N$ follows the Thomas distribution. We suppose that the random variable X follows the gamma distribution with characteristic function

$$\varphi_X(u) = \left(\frac{\mu}{\mu - iu}\right)^a. \tag{2.5.40}$$

From (2.5.24), (2.5.39) and (2.5.40) it follows that the characteristic function of the random sum $L = X_1 + X_2 + \cdots + X_K$ has the form

$$\varphi_L(u) = \exp\left\{\lambda\left[\left(\frac{\mu}{\mu - iu}\right)^a e^{\theta\left[\left(\frac{\mu}{\mu-iu}\right)^a - 1\right]} - 1\right]\right\}.$$

We suppose that the random variable N follows the Poisson distribution with probability generating function

$$P_N(z) = e^{\lambda(z-1)} \tag{2.5.41}$$

and the random variable V follows the renewal distribution corresponding to the distribution of the random variable D, which follows the Poisson distribution with probability generating function $P_D(z) = e^{\theta(z-1)}$. The probability generating function of the random variable V is

$$P_V(z) = \frac{1 - e^{\theta(z-1)}}{\theta(1-z)}. \tag{2.5.42}$$

From (2.5.41) and (2.5.42) it follows that the probability generating function $P_K(z) = P_N(P_V(z))$ of the random sum $K = V_1 + V_2 + \cdots + V_N$ has the form

$$P_K(z) = \exp\left\{\lambda\left[\frac{1 - e^{\theta(z-1)}}{\theta(1-z)} - 1\right]\right\}. \tag{2.5.43}$$

Hence the random sum $K = V_1 + V_2 + \cdots + V_N$ follows the Neyman type B distribution. We suppose that the random X follows the uniform distribution with characteristic function

$$\varphi_X(u) = \frac{e^{iu} - 1}{iu}. \tag{2.5.44}$$

From (2.5.24), (2.5.43) and (2.5.44) it follows that the characteristic function of the random sum $L = X_1 + X_1 + \cdots + X_K$ has the form

$$\varphi_L(u) = \exp\left\{ \lambda \left[\frac{1 - e^{\theta\left(\frac{e^{iu}-1}{iu} - 1\right)}}{\theta\left(1 - \frac{e^{iu}-1}{iu}\right)} - 1 \right] \right\}.$$

□

2.6 Recovery Time of a Partially Damaged System

We consider the occurrence time of a risk as the time point 0. The occurrence of the risk interrupts N operations of a system, where N is a discrete random variable with values in the set $\mathbf{N} = \{1, 2, \ldots\}$ and probability generating function $P_N(z)$.

We suppose that $\{X_n, n = 1, 2, \ldots\}$ is a sequence of continuous, positive, independent, and identically distributed random variables. The random variables of the sequence are distributed as the random variable X with distribution function $F_X(x)$.

The sequence of random variables $\{X_n, n = 1, 2, \ldots\}$ is independent of the random variable N.

If the random variable X_n denotes the time required for the recovery of the nth interrupted operation of the system then the random variable $T = \max(X_1, X_2, \ldots, X_N)$ denotes the time required for the recovery of the system.

The evaluation of the distribution function $F_T(t)$ of the random variable $T = \max(X_1, X_2, \ldots, X_N)$ is particularly important for the study of the behavior of the system after the occurrence of the risk. Since $F_T(t) = P(T \leq t)$ or equivalently

$$F_T(t) = P[\max(X_1, X_2, \ldots, X_N) \leq t] \tag{2.6.1}$$

then (2.6.1) implies that

$$F_T(t) = \sum_{n=1}^{\infty} P[\max(X_1, X_2, \ldots, X_N) \leq t | N = n] P(N = n)$$

or equivalently

$$F_T(t) = \sum_{n=1}^{\infty} P[\max(X_1, X_2, \ldots, X_n) \leq t | N = n] P(N = n). \qquad (2.6.2)$$

Since the event $[\max(X_1, X_2, \ldots, X_n) \leq t | N = n]$ implies the event $(X_1 \leq t, X_2 \leq t, \ldots, X_n \leq t | N = n)$ then (2.6.2) has the form

$$F_T(t) = \sum_{n=1}^{\infty} P(X_1 \leq t, X_2 \leq t, \ldots, X_n \leq t | N = n) P(N = n). \qquad (2.6.3)$$

The independence of the random variable N from the sequence of continuous, positive, independent, and identically distributed random variables $\{X_n, n = 1, 2, \ldots\}$ means the independence of the random variables $N, X_1, X_2, \ldots, X_n$.

Hence (2.6.3) has the form

$$F_T(t) = \sum_{n=1}^{\infty} P(X_1 \leq t, X_2 \leq t, \ldots, X_n \leq t) P(N = n). \qquad (2.6.4)$$

Since the random variables of the sequence $\{X_n, n = 1, 2, \ldots\}$ are independent then (2.6.4) has the form

$$F_T(t) = \sum_{n=1}^{\infty} P(X_1 \leq t) P(X_2 \leq t) \ldots P(X_n \leq t) P(N = n). \qquad (2.6.5)$$

Moreover, the assumption that the random variables of the sequence $\{X_n, n = 1, 2, \ldots\}$ are distributed as the random variable X having distribution function $F_X(x)$ then (2.6.5) has the form

$$F_T(t) = \sum_{n=1}^{\infty} F_X^n(t) P(N = n). \qquad (2.6.6)$$

From (2.6.6) it follows that the distribution function of the random variable $T = \max(X_1, X_2, \ldots, X_N)$ is

$$F_T(t) = P_N(F_X(t)), \; t > 0. \qquad (2.6.7)$$

An interesting special case of (2.6.7) arises if the random variables of the sequence $\{X_n, n = 1, 2, \ldots\}$ follow the exponential distribution with distribution function

$$F_X(x) = 1 - e^{-\mu x}, \; x > 0, \; \mu > 0 \qquad (2.6.8)$$

and the random variable N follows the Sibuya distribution with probability generating function

$$P_N(z) = 1 - (1-z)^\gamma,\ 0 < \gamma \le 1. \tag{2.6.9}$$

From (2.6.7), (2.6.8) and (2.6.9) it follows that the distribution function of the random variable $T = \max(X_1, X_2, \ldots, X_N)$ has the form $F_T(t) = 1 - [1-(1-e^{-\mu t})]^\gamma$, $t > 0$ or equivalently the form $F_T(t) = 1 - e^{-\mu\gamma t}$, $t > 0$.

Hence, in this special case, the random variable $T = \max(X_1, X_2, \ldots, X_N)$ follows the exponential distribution with parameter $\mu\gamma$.

Another special case of (2.6.7) arises if the random variables of the sequence $\{X_n, n = 1, 2, \ldots\}$ follow the uniform distribution with distribution function

$$F_X(x) = x\ ,\ 0 < x < 1, \tag{2.6.10}$$

and the random variable N follows the geometric type II distribution with probability generating function

$$P_N(z) = \frac{pz}{1-qz}. \tag{2.6.11}$$

From (2.6.7), (2.6.10) and (2.6.11) it follows that the distribution function of the random variable $T = \max(X_1, X_2, \ldots, X_N)$ has the form $F_T(t) = \frac{pt}{1-qt},\ \ 0 < t < 1$.

The present section is based on the assumption that the risk occurrence, interrupting N operations of a system, is realized at given time point called time point 0. This assumption does not agree with the random character of risk management. In practice, the time point 0 is a sufficiently small and closed time interval where a risk occurs with probability 1. Consequently, any point of such an interval can be considered as the occurrence time point of a risk and the random variable $T = \max(X_1, X_2, \ldots, X_N)$ can the be interpreted as a stochastic model for the recovery time of a partially damaged system.

The distribution function $F_T(t) = P_N(F_X(t))$ of the random variable $T = \max(X_1, X_2, \ldots, X_N)$ provides probabilistic information which makes the time interval $[0, T]$ particularly important for the implementation of the risk management principles and operations.

2.7 Time of First Damage of a System Threatened by a Random Number of Risks

The discrete random variable N with values in the set $\mathbf{N} = \{1, 2, \ldots\}$ and probability generating function $P_N(z)$ is independent of the sequence of continuous, positive, independent, and identically distributed random variables $\{X_n, n = 1, 2, \ldots\}$.

The random variables of the above sequence are distributed as the random variable X having distribution function $F_X(x)$.

If the random variable N denotes the number of risks threatening a system at the time point 0 and the random variable X_n denotes the occurrence time of the nth risk then the random variable $T = \min(X_1, X_2, \ldots, X_N)$ denotes the time of the first risk occurrence. The consideration of a system under a random number N of independent competing risks means the use of the random variable $T = \min(X_1, X_2, \ldots, X_N)$ as a fundamental stochastic model for investigating the evolution of this system.

The evaluation of the distribution function $F_T(t)$ of the random variable $T = \min(X_1, X_2, \ldots, X_N)$ is very important for the consideration of a system under a random number of independent competing risks. We have $F_T(t) = P(T \leq t)$ or equivalently

$$F_T(t) = P[\min(X_1, X_2, \ldots, X_N) \leq t]. \tag{2.7.1}$$

Since the event $[\min(X_1, X_2, \ldots, X_N) > t]$ is the complement of the event $[\min(X_1, X_2, \ldots, X_N) \leq t]$ then (2.7.1) implies that $F_T(t) = 1 - P[\min(X_1, X_2, \ldots, X_N) > t]$ or equivalently

$$F_T(t) = 1 - \sum_{n=1}^{\infty} P(\min(X_1, X_2, \ldots, X_N) > t | N = n) P(N = n). \tag{2.7.2}$$

From (2.7.2) it follows that

$$F_T(t) = 1 - \sum_{n=1}^{\infty} P(\min(X_1, X_2, \ldots, X_n) > t | N = n) P(N = n). \tag{2.7.3}$$

Since the event $[\min(X_1, X_2, \ldots, X_n) > t | N = n]$ implies the event $(X_1 > t, X_2 > t, \ldots, X_n > t | N = n)$ then (2.7.3) has the form

$$F_T(t) = 1 - \sum_{n=1}^{\infty} P(X_1 > t, X_2 > t, \ldots, X_n > t | N = n) P(N = n). \tag{2.7.4}$$

The independence of the random variable N and the sequence of continuous, positive, independent, and identically distributed random variables $\{X_n, n = 1, 2, \ldots\}$ means the independence of the random variables $N, X_1, X_2, \ldots, X_n$.

Hence (2.7.4) has the form

$$F_T(t) = 1 - \sum_{n=1}^{\infty} P(X_1 > t, X_2 > t, \ldots, X_n > t) P(N = n). \tag{2.7.5}$$

Since the random variables of the sequence $\{X_n, n = 1, 2, \ldots\}$ are independent then (2.7.5) has the form $F_T(t) = 1 - \sum_{n=1}^{\infty} P(X_1 > t)P(X_2 > t)\ldots P(X_n > t)$ $P(N = n)$ or equivalently the form

$$F_T(t) = 1 - \sum_{n=1}^{\infty} [1 - P(X_1 \leq t)]\,[1 - P(X_2 \leq t)]\ldots[1 - P(X_n \leq t) \leq t]P(N = n). \tag{2.7.6}$$

Moreover, the assumption that the random variables of the sequence $\{X_n, n = 1, 2, \ldots\}$ are distributed as the random variable X having distribution function $F_X(x)$ implies that (2.7.6) has the form

$$F_T(t) = 1 - \sum_{n=1}^{\infty} [1 - F_X(t)]^n P(N = n). \tag{2.7.7}$$

From (2.7.7) it follows that the distribution function of the random variable $T = \min(X_1, X_2, \ldots, X_N)$ has the form

$$F_T(t) = 1 - P_N(1 - F_X(t)),\ t > 0. \tag{2.7.8}$$

An interesting special case of (2.7.8) arises if the random variables of the sequence $\{X_n, n = 1, 2, \ldots\}$ follow the beta distribution with parameters $\alpha, 1$ or equivalently the distribution function of the random variables of the above sequence has the form

$$F_X(x) = x^{\alpha},\ 0 < x < 1 \tag{2.7.9}$$

and the random variable N follows the Sibuya distribution with probability generating function

$$P_N(z) = 1 - (1 - z)^{\gamma},\ 0 < \gamma \leq 1. \tag{2.7.10}$$

From (2.7.8), (2.7.9) and (2.7.10) it follows that the distribution function of the random variable $T = \min(X_1, X_2, \ldots, X_N)$ has the form $F_T(t) = 1 - \{1 - [1 - (1 - t^{a})]^{\gamma}\}$ or equivalently the form $F_T(t) = t^{\alpha\gamma}, 0 < t < 1$.

Hence, in this case, the random variable $T = \min(X_1, X_2, \ldots, X_N)$ follows the beta distribution with parameters $\alpha\gamma, 1$.

The role of the random variable $T = \min(X_1, X_2, \ldots, X_N)$ in the consideration of a system under a random number N of independent and competing risks becomes very important if the occurrence of one of these risks implies the destruction of the system. In this case the random variable $T = \min(X_1, X_2, \ldots, X_N)$ denotes the life time of the system.

2.8 Time of First Major Damage

We consider the sequence of continuous, positive, independent, and identically distributed random variables $\{C_n, n = 1, 2, \ldots\}$. The random variables of the sequence are distributed as the random variable C having characteristic function

$$\varphi_C(u). \tag{2.8.1}$$

The random variable $C_n, n = 1, 2, \ldots$ denotes the time between the $(n-1)$th and the nth occurrence of a risk

We consider the sequence of continuous, positive, independent, and identically distributed random variables $\{X_n, n = 1, 2, \ldots\}$.

The random variables of the sequence are distributed as the random variable X having distribution function

$$F_X(x). \tag{2.8.2}$$

The random variable X_n denotes the size of the damage from the nth occurrence of the risk.

Let θ be a positive real number. If $X_n > \theta$ then the damage due to the nth occurrence of the risk is considered as major one. If p is the probability of the event that the damage due to the nth occurrence of the risk is a major one then $p = P(X_n > \theta)$ or equivalently $p = 1 - P(X_n \leq \theta)$.

Hence (2.8.2) implies that $p = 1 - F_X(\theta)$.

Let N be a random variable denoting the number of risk occurrences required to get the first major damage. The random variable N follows the geometric type II distribution with probability function $P(N = n) = pq^{n-1},\ q = 1 - p,\ n = 1, 2, \ldots$ and probability generating function

$$P_N(z) = \frac{pz}{1 - qz}. \tag{2.8.3}$$

Moreover, the random variable N is independent of the sequence of continuous, positive, independent, and identically distributed random variables $\{C_n, n = 1, 2, \ldots\}$.

The random sum $Y = C_1 + C_2 + \cdots + C_N$ denotes the occurrence time of the first major damage. From Sect. (2.5), (2.8.1), and (2.8.3) it follows that the characteristic function of the random sum $Y = C_1 + C_2 + \cdots + C_N$ is

$$\varphi_Y(u) = \frac{p\varphi_C(u)}{1 - q\varphi_C(u)}. \tag{2.8.4}$$

A special case of the characteristic function $\varphi_Y(u)$ arises if the random variables of the sequence $\{C_n, n = 1, 2, \ldots\}$ follow the exponential distribution with characteristic function

$$\varphi_C(u) = \frac{\mu}{\mu - iu}, \; \mu > 0. \tag{2.8.5}$$

From (2.8.4) and (2.8.5) it follows that $\varphi_Y(u) = \frac{v}{v - iu}$ where $v = p\mu$.

Hence, in this case, the random sum $Y = C_1 + C_2 + \cdots + C_N$ follows the exponential distribution with parameter $v = p\mu$.

The time $Y = C_1 + C_2 + \cdots + C_N$, of the first major damage from the occurrence of a risk, is particularly significant in practice if the realization of the first major damage implies the destruction of the organization threatened by the risk. In that case the random sum $Y = C_1 + C_2 + \cdots + C_N$ denotes the life time of the organization. A direct consequence of that case is the recognition of the importance of the random sum $Y = C_1 + C_2 + \cdots + C_N$ and the corresponding characteristic function $\varphi_Y(u) = \frac{p\varphi_C(u)}{1 - q\varphi_C(u)}$ in the formulation and investigation of stochastic models describing the behavior and evolution of an organization. The form of the characteristic function $\varphi_C(u)$ reflects the difficulty of investigating of such stochastic models. The presence of the time value of money in stochastic models, having as a constituent element the random sum $Y = C_1 + C_2 + \cdots + C_N$, and describing the evolution of a system, is of significant practical importance.

2.9 Number of Ongoing Risk Occurrences

A risk occurrence is considered as an ongoing one, at a given time point, if the risk cause is active at that time point. The present section concentrates on the establishment of an application of a result of service systems theory for evaluating the distribution of the random variable denoting the ongoing occurrences of a risk. The presentation of that application is based on the concept of ordered sample of continuous, independent, and identically distributed random variables, and a property of the homogeneous Poisson process.

Let $C_1, C_2, \ldots, C_n$ be random variables. The random variables $L_1, L_2, \ldots, L_n$ is an ordered random sample corresponding to the random variables $C_1, C_2, \ldots, C_n$ if the random variable $L_\kappa, \kappa = 1, 2, \ldots, n$ denotes the κth smallest value among $C_1, C_2, \ldots, C_n$.

We suppose that the random variables $C_1, C_2, \ldots, C_n$ are continuous, independent, and identically distributed. Moreover, we suppose that the random variables $C_1, C_2, \ldots, C_n$ are equally distributed with the random variable C having probability generating function $f_C(c)$.

In this case the joint probability density function of the ordered random sample $L_1, L_2, \ldots, L_n$ has the form

$$f_{L_1,L_2,\ldots,L_n}(l_1, l_2, \ldots, l_n) = n!\, f_C(l_1) f_C(l_2) \ldots f_C(l_n). \tag{2.9.1}$$

We consider the homogeneous Poisson process $\{N(t), t \geq 0\}$ with $E(N(t)) = \lambda t$.

The following theorem establishes the property of homogeneous Poisson process $\{N(t), t \geq 0\}$ which constitutes the structural element of the present section.

Theorem 2.9.1 *Let $\{N(t), t \geq 0\}$ be a homogeneous Poisson process with $E(N(t)) = \lambda t$ and $W_\kappa, \kappa = 1, 2, \ldots$ a random variable denoting the waiting time for the occurrence of the κth event of the homogeneous Poisson process $\{N(t), t \geq 0\}$.*

If $g(w_1, w_2, \ldots, w_n | N(t) = n)$ is the joint probability density function of the random variable $W_1, W_2, \ldots, W_n$ when $N(t) = n$ then $g(w_1, w_2, \ldots, w_n | N(t) = n) = \frac{n!}{t^n}$, $0 < w_1 < w_2 < \ldots < w_n < t$.

Proof We consider the time points $t_1, t_2, \ldots, t_n, t_{n+1}$ satisfying $t_1 < t_2 < \ldots < t_n < t_{n+1}$ and $t = t_{n+1}$.

Moreover, we consider the positive real numbers $h_1, h_2, \ldots, h_n$ satisfying $t_1 + h_1 < t_2, \ldots, t_n + h_n < t_{n+1}$.

We have

$$P\{[t_1 \leq W_1 \leq t_1 + h_1, \ldots, t_n \leq W_n \leq t_n + h_n] | (N(t) = n)\} = \frac{P\{[t_1 \leq W_1 \leq t_1 + h_1, \ldots, t_n \leq W_n \leq t_n + h_n], (N(t) = n)\}}{P(N(t) = n)}. \tag{2.9.2}$$

Since the event $\{[t_1 \leq W_1 \leq t_1 + h_1, \ldots, t_n \leq W_n \leq t_n + h_n], (N(t) = n)\}$ is equivalent to the event $\{[N(t_1 + h_1) - N(t_1) = 1, \ldots, N(t_n + h_n) - N(t_n) = 1], (N(t) = n)\}$ then (2.9.2) implies that

$$P\{[t_1 \leq W_1 \leq t_1 + h_1, \ldots, t_n \leq W_n \leq t_n + h_n] | (N(t) = n)\} = \frac{P\{[N(t_1 + h_1) - N(t_1) = 1, \ldots, N(t_n + h_n) - N(t_n) = 1], (N(t) = n)\}}{P(N(t) = n)}. \tag{2.9.3}$$

Since the event $\{[N(t_1 + h_1) - N(t_1) = 1, \ldots, N(t_n + h_n) - N(t_n) = 1], (N(t) = n)\}$ is equivalent to the event $\{[N(t_1 + h_1) - N(t_1) = 1, \ldots, N(t_n + h_n) - N(t_n) = 1, 0 \text{ } \textit{no events elsewhere in } [0, t]]\}$ then (2.9.3) implies that

$$P\{[t_1 \leq W_1 \leq t_1 + h_1, \ldots, t_n \leq W_n \leq t_n + h_n] | (N(t) = n)\} =$$

$$\frac{P\{[N(t_1 + h_1) - N(t_1) = 1, \ldots, N(t_n + h_n) - N(t_n) = 1, \textit{ no events elsewhere in } [0, t]]\}}{P(N(t) = n)}. \tag{2.9.4}$$

Since the process $\{N(t), t \geq 0\}$ is a homogeneous Poisson process with $E(N(t)) = \lambda t$ then (2.9.4) implies that

$$P\{[t_1 \leq W_1 \leq t_1 + h_1, \ldots, t_n \leq W_n \leq t_n + h_n]|(N(t) = n)\}$$
$$= \frac{\lambda h_1 e^{-\lambda h_1} \ldots e^{-\lambda(t-h_1-\ldots-h_n)}}{e^{-\lambda t}\frac{(\lambda t)^n}{n!}}$$

or equivalently

$$P\{[t_1 \leq W_1 \leq t_1 + h_1, \ldots, t_n \leq W_n \leq t_n + h_n]|(N(t) = n)\} = \frac{n!}{t^n} h_1 \ldots h_n. \quad (2.9.5)$$

From (2.9.5) it follows that

$$\frac{P\{[t_1 \leq W_1 \leq t_1 + h_1, \ldots, t_n \leq W_n \leq t_n + h_n]|(N(t) = n)\}}{h_1 h_2 \ldots h_n} = \frac{n!}{t^n}. \quad (2.9.6)$$

Hence the joint probability generating function of the random variables $W_1, W_2, \ldots, W_n$ given that $N(t) = n$ is

$$g(w_1, w_2, \ldots, w_n|N(t) = n) =$$
$$\lim \frac{P\{[t_1 \leq W_1 \leq t_1 + h_1, \ldots, t_n \leq W_n \leq t_n + h_n]|(N(t) = n)\}}{h_1 \ldots h_n}$$
$$h_1 \to 0, \ldots, h_n \to 0 \quad (2.9.7)$$

From (2.9.6) and (2.9.7) it follows that $g(w_1, w_2, \ldots, w_n|N(t) = n) = \lim\limits_{h_1 \to 0,\, h_2 \to 0, \ldots,\, h_n \to 0} \frac{n!}{t^n}$, or equivalently

$$g(w_1, w_2, \ldots, w_n|N(t) = n) = \frac{n!}{t^n}. \quad (2.9.8)$$

From Theorem 2.9.1 and (2.9.8) we get the following conclusion. If $N(t) = n$, that is n events of the homogeneous Poisson process have occurred in the time interval $[0, t]$ and the continuous, independent, positive, and identically distributed random variables $V_1, V_2, \ldots, V_n$ represent the unordered occurrence times of these events then the random variables $W_1, W_2, \ldots, W_n$ is the ordered sample of the random variables $V_1, V_2, \ldots, V_n$.

From (2.9.1) and Theorem 2.9.1 it follows that the random variables $V_1, V_2, \ldots, V_n$ are equally distributed with the random variable V which follows the uniform distribution with probability density function $f_V(v) = \frac{1}{t}, \ 0 < v < t$.

The following result constitutes an application in risk management of a result of service systems theory. □

Theorem 2.9.2 *Let* $\{N(t), t \geq 0\}$ *be a homogeneous Poisson process with* $E(N(t)) = \lambda t$.

We suppose that the random variable $N(t)$ *denotes the frequency of a risk in the time interval* $[0, t]$ *and* $\{Y_n, n = 1, 2, \ldots\}$ *is a sequence of continuous, positive,* *independent, and identically distributed random variables. The random variables of the sequence represent the durations of the risk occurrences. Moreover, these random variables are equally distributed with the random variable Y having distribution function $F_Y(y)$.*

If the random variable $\Pi(t)$ denotes the number of the ongoing risk occurrences at the time point t then the probability generating function of the random variable $\Pi(t)$ is $P_{\Pi(t)}(z) = e^{\lambda pt(z-1)}$ where

$$p = \int_0^t \frac{1 - F_Y(y)}{t} dy.$$

Proof We have $P_{\Pi(t)}(z) = E\left(z^{\Pi(t)}\right)$ or equivalently

$$P_{\Pi(t)}(z) = E\left(E\left(z^{\Pi(t)} | N(t)\right)\right). \tag{2.9.9}$$

From (2.9.9) it follows that

$$P_{\Pi(t)}(z) = \sum_{n=0}^{\infty} E\left(z^{\Pi(t)} | N(t) = n\right) P(N(t) = n). \tag{2.9.10}$$

Since

$$P(N(t) = n) = e^{-\lambda t} \frac{(\lambda t)^n}{n!}$$

then (2.9.10) has the form

$$P_{\Pi(t)}(z) = \sum_{n=0}^{\infty} E\left(z^{\Pi(t)} | N(t) = n\right) e^{-\lambda t} \frac{(\lambda t)^n}{n!}. \tag{2.9.11}$$

We suppose that v is the time point of a risk occurrence where $0 < v < t$.

Since the continuous and positive random variable Y denotes the duration of the risk occurrence arising at the time point v then the probability of the event that this risk occurrence will be an ongoing one at the time point t is $P(Y > t - v)$.

Since $P(Y > t - v) = 1 - P(Y \leq t - v)$ and $F_Y(y)$ is the distribution function of the random variable Y then

$$P(Y > t - v) = 1 - F_Y(t - v). \tag{2.9.12}$$

If $N = n$, that is n risk occurrences arise in the time interval $[0, t]$, then Theorem 2.9.1 implies that the unordered time points of the n risk occurrences in the time interval $[0, t]$ are continuous, independent random variables $V_1, V_2, \ldots, V_n$ equally distributed with the random variable V which follows the uniform distribution with probability density function $f_V(v) = \frac{1}{t},\ 0 < v < t$.

Hence the probability of the event that a risk occurrence arising in the time interval $[0, t]$ is ongoing at the time point t independently of the other risk occurrences in the time interval $[0, t]$, according to (2.9.12), has the form

$$p = \int_0^t \frac{1 - F_Y(t - v)}{t} dv$$

or equivalently the form

$$p = \int_0^t \frac{1 - F_Y(y)}{t} dy.$$

In this case if $N(t) = n$, that is n risk occurrences arise in the time interval $[0, t]$ then the random variable $\Pi(t)|N(t) = n$, which denotes the number of risk occurrences in the time interval $[0, t]$ and which risk occurrences are ongoing at the time point t, follows the binomial distribution with parameters n and p.

Hence the probability generating function of the random variable $\Pi(t)|N(t) = n$ is

$$E\left(z^{\Pi(t)}|N(t) = n\right) = (pz + q)^n \tag{2.9.13}$$

where $q = 1 - p$.

From (2.9.11) and (2.9.13) it follows that the probability generating function $P_{\Pi(t)}(z)$ of the random variable $\Pi(t)$ has the form

$$P_{\Pi(t)}(z) = \sum_{n=0}^{\infty} (pz + q)^n e^{-\lambda t} \frac{(\lambda t)^n}{n!}$$

or equivalently the form

$$P_{\Pi(t)}(z) = e^{-\lambda t} \sum_{n=0}^{\infty} \frac{[\lambda t(pz + q)]^n}{n!}. \tag{2.9.14}$$

From (2.9.14) it follows that $P_{\Pi(t)}(z) = e^{-\lambda t} e^{\lambda t(pz+q)}$ or equivalently $P_{\Pi(t)}(z) = e^{\lambda pt(z-1)}$.

Hence the random variable $\Pi(t)$ follows the Poisson distribution with parameter λpt.

Since the random variable $\Pi(t)$ denotes the number of the occurrences of a risk in the time interval $[0, t]$ and which occurrences are ongoing at the time point t then this random variable can be used in formulating stochastic models for describing operations suitable for financing damages due to the occurrences of that risk. The constituent elements of the contribution of the present section are the following two. The first constituent element is the introduction of the concept of ongoing risk occurrence. The second element is the application of a significant result of service systems theory in modeling the concept of the number of ongoing occurrences of a risk. That application provides risk managers with the ability to get a holistic consideration of the number of ongoing occurrences of a risk. The fundamental assumption of the proposed application is that the frequency of the risk in the time interval $[0, t]$ is represented by a homogeneous Poisson $\{N(t), t \geq 0\}$.

This assumption does not restrict the significance of the proposed application since the homogeneous Poisson process is considered as a very efficient model of the frequency of a risk in the time interval $[0, t]$.

The conclusion that the number of ongoing occurrences of a risk at a given time point t is represented by the random variable $\Pi(t)$ following the Poisson distribution with parameter λpt can be considered as a very good reason for modeling the number of ongoing risk occurrences at a random time point. □

2.10 Multiplicative Models of Risk Severity

The purpose of the present section is the formulation and investigation of a stochastic multiplicative model for the description and investigation of risk severity. The model is based on the product of two continuous, positive, and independent random variables.

We suppose that the duration of a risk is represented by the continuous, and positive random variable S with distribution function $F_S(s)$, probability density function $f_S(s)$, and characteristic function $\varphi_S(u)$.

We suppose that U is a continuous and positive random variable with distribution function $F_U(\upsilon)$, probability density function $f_U(\upsilon)$, and characteristic function $\varphi_U(u)$.

The random variable U denotes the damage, per unit of time, due to the occurrence of a risk. We suppose that the random variable S is independent of the random variable U.

The random variable $X = SU$ represents the severity of risk. The independence of random variables S, U permits the evaluation of the distribution function $F_X(x)$, the evaluation of the probability density function $f_X(x)$, and the evaluation of the characteristic function $\varphi_X(u)$ of the random variable $X = SU$.

We have $F_X(x) = P(X \leq x)$ or equivalently $F_X(x) = P(SU \leq x)$.

Hence

$$F_X(x) = \int_0^\infty P(SU \leq x | U = \upsilon) f_U(\upsilon) d\upsilon$$

or equivalently

$$F_X(x) = \int_0^\infty P(vS \le x | U = v) f_U(v) dv. \tag{2.10.1}$$

Since the random variable S is independent of the random variable U then (2.10.1) has the form

$$F_X(x) = \int_0^\infty P(vS \le x) f_U(v) dv$$

or equivalently the form

$$F_X(x) = \int_0^\infty P\left(S \le \frac{x}{v}\right) f_U(v) dv. \tag{2.10.2}$$

Since

$$P\left(S \le \frac{x}{v}\right) = F_S\left(\frac{x}{v}\right)$$

then (2.10.2) implies that the distribution function $F_X(x)$ of the random variable $X - SU$ is

$$F_X(x) = \int_0^\infty F_S\left(\frac{x}{v}\right) f_U(v) dv. \tag{2.10.3}$$

It is obvious that the following formula is also valid

$$F_X(x) = \int_0^\infty F_U\left(\frac{x}{s}\right) f_S(s) ds. \tag{2.10.4}$$

From (2.10.3) and (2.10.4) it follows that the probability density function of the random variable $X = SU$ has the form

$$f_X(x) = \int_0^\infty \frac{1}{v} f_S\left(\frac{x}{v}\right) f_U(v) dv$$

or equivalently the form

$$f_X(x) = \int_0^\infty \frac{1}{s} f_U\left(\frac{x}{s}\right) f_S(s) ds.$$

If $\varphi_X(u)$ is the characteristic function of the random variable $X = SU$ then $\varphi_X(u) = E(e^{iuX})$ or equivalently $\varphi_X(u) = E(e^{iuSU})$.
Hence

$$\varphi_X(u) = \int_0^\infty E(e^{iuSU} | U = \upsilon) f_U(\upsilon) d\upsilon$$

or equivalently

$$\varphi_X(u) = \int_0^\infty E(e^{iu\upsilon S} | U = \upsilon) f_U(\upsilon) d\upsilon. \tag{2.10.5}$$

Since the random variable S is independent of the random variable U then (2.10.5) has the form

$$\varphi_X(u) = \int_0^\infty E(e^{iu\upsilon S}) f_U(\upsilon) d\upsilon. \tag{2.10.6}$$

Since $E(e^{iu\upsilon S}) = \varphi_S(u\upsilon)$ then (2.10.6) implies that the characteristic function $\varphi_X(u)$ of the random variable $X = SU$ is

$$\varphi_X(u) = \int_0^\infty \phi_S(u\upsilon) f_U(\upsilon) d\upsilon \tag{2.10.7}$$

It is obvious that the following formula is also valid

$$\varphi_X(u) = \int_0^\infty \phi_U(us) f_S(s) ds$$

The consideration of special cases of the distribution of the stochastic model $X = SU$, with use of the corresponding characteristic function $\varphi_X(u)$, is very significant for the practical applications of the stochastic model $X = SU$ in the description and analysis of concepts and operations of risk management.

Special cases of the distribution of the stochastic model $X = SU$, having probability distribution functions with unique mode a the point 0, are of particular practical importance.

The probability density function $f_\Pi(\Pi)$ of the continuous random variable Π is said unimodal at the point 0 if this probability density function has a unique maximum at the point 0. The establishment of the property of unimodality at the point 0 for the probability density function $f_X(x)$ of risk severity $X = SU$ is based on a result of Khintchine and a result of Medgyessy. The result of Khintchine states that the probability density function $f_\Pi(\pi)$ is unimodal at the point 0 if the corresponding characteristic function $\varphi_\Pi(u)$ has the form

$$\varphi_\Pi(u) = \int_0^1 \varphi_D(uy)dy \tag{2.10.8}$$

where $\varphi_D(u)$ is the characteristic function of a continuous random variable D.

The result of Medgyessy states that if the continuous random variable Π has probability generating function $f_\Pi(\pi)$ which is unimodal at the point 0 and B is a continuous random variable independent of the random variable Π then the random variable ΠB has a probability density function with a unique mode at the point 0.

If the random variable U follows the uniform distribution with probability density function $f_U(\upsilon) = 1,\ 0 < \upsilon < 1$ then (2.10.7) implies that the characteristic function of risk severity $X = SU$ has the form

$$\varphi_X(u) = \int_0^1 \varphi_S(u\upsilon)d\upsilon. \tag{2.10.9}$$

From (2.10.8) and (2.10.9) it follows that the probability density function

$$f_X(x) = \int_0^1 \frac{1}{\upsilon} f_S\left(\frac{x}{\upsilon}\right) d\upsilon$$

of risk severity $X = SU$ is unimodal at the point 0.

If the continuous and positive random variable S has probability density function $f_S(s)$ with unique mode at the point 0 or the continuous and positive random variable U has probability density function $f_U(\upsilon)$ with unique mode at the point 0 then the result of Medgyessy implies that the random variable $X = SU$ has probability density function

$$f_X(x) = \int_0^\infty \frac{1}{\upsilon} f_S\left(\frac{x}{\upsilon}\right) f_U(\upsilon) d\upsilon,$$

or equivalently probability density function

$$f_X(x) = \int_0^\infty \frac{1}{s} f_U\left(\frac{x}{s}\right) f_S(s) ds$$

which has a unique mode at the point 0. The existence of a unique mode at the point 0 for the probability density function $f_X(x)$ of the random variable $X = SU$ implies that the event the size of the damage, due to an occurrence of the risk, to be in an area right to the point 0 has a significant probability. That means that the organization threatened by the risk can select the retention of the risk instead of the transfer of the risk.

Significant theoretical and practical interest has the special case of the stochastic multiplicative model $X = SU$ if the continuous and positive random variable S follows the exponential distribution with characteristic function

$$\varphi_S(u) = \frac{\mu}{\mu - iu}. \tag{2.10.10}$$

From (2.10.7) and (2.10.10) it follows that the characteristic function of the stochastic multiplicative model $X = SU$ has the form

$$\varphi_X(u) = \int_0^\infty \frac{\mu}{\mu - iuv} f_U(v) dv.$$

Since the probability density function $f_S(s) = \mu e^{-\mu s}$ has a unique mode at the point 0 then the probability density function

$$f_X(x) = \int_0^\infty \frac{\mu}{v} e^{-\mu \frac{x}{v}} f_U(v) dv$$

corresponding to the characteristic function

$$\varphi_X(u) = \int_0^\infty \frac{\mu}{\mu - iuv} f_U(v) dv$$

has a unique mode at the point 0. Characteristic functions of the form

$$\varphi_X(u) = \int_0^\infty \frac{\mu}{\mu - iuv} f_U(v) dv$$

belong to the class of infinitely divisible characteristic functions having important applications to stochastic processes.

From a theoretical and practical point of view, it is of particular interest to investigate properties of the probability density function $f_U(\upsilon)$ of the random variable U which are transferred to the probability density function

$$f_X(x) = \int_0^\infty \frac{\mu}{\upsilon} e^{-\mu\frac{x}{\upsilon}} f_U(\upsilon) d\upsilon$$

of the random variable $X = SU$.

An interpretation of the stochastic multiplicative model $X = SU$ in the area of fundamental risk control operations is the following.

If the continuous and positive random variable U takes values in the interval $(0, 1)$ then the random variable U can be considered as a coefficient describing the impact of a risk control operation. In this case the presence of the random variable U and the presence of the random variable S in the stochastic multiplicative model $X = SU$ permit the interpretation of the random variable X as the consequence of the application of a risk control operation. This interpretation of the stochastic multiplicative model $X = SU$ in the area of fundamental risk control operations requires the consideration of the continuous and positive random variable S as a positive component of the concept of risk.

Since the random variable S is continuous and positive then this random variable can represent the severity or the duration of a risk. Hence the stochastic multiplicative model $X = SU$ can be used for the description and analysis of risk severity and risk duration reduction operations. Such applications of the stochastic multiplicative model $X = SU$ constitute the main purpose of the third chapter of the present work.

Particular practical interest has the stochastic multiplicative model $X = SU$ with the continuous and positive random variable S having the form of a random sum. In this case the investigation of the stochastic multiplicative model $X = SU$ is based on the corresponding characteristic function $\varphi_X(u)$.

2.11 Riskiness

We consider a risk with frequency denoted by the discrete random variable N taking values in the set $\mathbf{N}_0 = \{0, 1, 2, \ldots\}$, severity denoted by the continuous and positive random variable X, and duration denoted by the continuous and positive random variable S.

The continuous and positive random variable $R = NXS$ is said riskiness of the risk or simply riskiness. Since riskiness $R = NXS$ is proportional to risk frequency N, risk severity X, and risk duration S then riskiness can be used as a model for describing the aversion of a person for a risk with frequency N, severity X, and duration S.

The distribution function $F_R(r)$, the probability density function $f_R(r)$, and the characteristic function $\varphi_R(u)$ of riskiness $R = NXS$ are the analytical tools implementing the theoretical and practical applicability of that concept.

The present section concentrates on the implementation of two purposes. The first purpose is the evaluation of the characteristic function $\varphi_R(u)$ of riskiness $R = NXS$.

The choice for evaluating the characteristic function $\varphi_R(u)$ is based on the important applications of the results of the theory of characteristic functions. The evaluation of the characteristic function $\varphi_R(u)$ of riskiness $R = NXS$ is based on the independence of the random variables N, X, S.

The second purpose is the establishment of the unimodality at the point 0 of the probability density function of riskiness $R = NXS$ by making use of the independence of the random variables N, X, S, the unimodality at the point 0 of the probability density function of the random variable X or equivalently the unimodality at the point 0 of the probability density function of the random variable S, the integral representation of a characteristic function corresponding to a probability density function with unique mode at the point 0, and the probability density function of the product of two independent and continuous random variables one of which has a probability density function with unique mode at the point 0.

The significance of the purposes of the present section is based on the presence of the fundamental quantitative components of risk, that is risk frequency N, risk severity X, and risk duration S in the definition of riskiness $R = NXS$.

The present section makes quite clear that the characteristic function $\varphi_R(u)$ constitutes a strong analytical tool for investigating the probabilistic behavior of riskiness. The establishment of a sufficient condition for evaluating the characteristic function of riskiness is provided by the following theorem.

Theorem 2.11.1 *We suppose that N is a discrete random variable with values in the set $\mathbf{N}_0 = \{0, 1, 2, \ldots\}$ and probability function $P(N = n) = p_n$, $n = 0, 1, 2, \ldots$, X is a continuous and positive random variable with probability density function $f_X(x)$ and characteristic function $\varphi_X(u)$, and S is a continuous and positive random variable with probability density function $f_S(s)$ and characteristic function $\varphi_S(u)$. If the random variables N, X, S are independent then the characteristic function of the random variable $R = NXS$ has the form*

$$\varphi_R(u) = \sum_{n=0}^{\infty} p_n \int_0^{\infty} \varphi_X(nus) f_S(s) ds$$

or equivalently the form

$$\varphi_R(u) = \sum_{n=0}^{\infty} p_n \int_0^{\infty} \varphi_S(nux) f_X(x) dx.$$

Proof The independence of the random variables N, X, S implies the independence of the random variables X, S.

We consider the random variable XS with characteristic function $\varphi_{XS}(u)$. We have $\varphi_{XS}(u) = E(E(e^{iuXS}|S))$ or equivalently

$$\varphi_{XS}(u) = \int_0^\infty E(e^{iuXS}|S = s)f_S(s)ds. \tag{2.11.1}$$

From (2.11.1) it follows that

$$\varphi_{XS}(u) = \int_0^\infty E(e^{iusX}|S = s)f_S(s)ds. \tag{2.11.2}$$

Since the random variable X is independent of the random variable S then (2.11.2) implies that

$$\varphi_{XS}(u) = \int_0^\infty E(e^{iusX})f_S(s)ds. \tag{2.11.3}$$

Since

$$E(e^{iusX}) = \varphi_X(us) \tag{2.11.4}$$

then (2.11.3) and (2.11.4) imply that the characteristic function of the random variable XS is

$$\varphi_{XS}(u) = \int_0^\infty \varphi_X(us)f_S(s)ds. \tag{2.11.5}$$

It is easily seen that (2.11.5) has the equivalent form

$$\varphi_{XS}(u) = \int_0^\infty \varphi_S(ux)f_X(x)dx. \tag{2.11.6}$$

The characteristic function $\varphi_{XS}(u)$ in (2.11.5) or (2.11.6) of the random variable XS and the proof of independence of the random variable N and the random variable XS are required for the evaluation of the characteristic function $\varphi_R(u)$ of riskiness $R = NXS$.

Let $\varphi_N(\xi)$ be the characteristic function of the random variable N, $\varphi_{XS}(u)$ be the characteristic function of the random variable XS and $\varphi_{N,XS}(\xi, u)$ be the characteristic function of the vector (N, XS) of random variables N, XS.

The proof of independence of the random variables N, XS requires the proof of the relationship

$$\varphi_{N,XS}(\xi, u) = \varphi_N(\xi)\varphi_{XS}(u). \tag{2.11.7}$$

We have

$$\varphi_{N,XS}(\xi, u) = E\big(e^{i\xi N + iuXS}\big).$$

or equivalently

$$\varphi_{N,XS}(\xi, u) = E\big(E\big(e^{i\xi N + iuXS}|S\big)\big). \tag{2.11.8}$$

From (2.11.8) it follows that

$$\varphi_{N,XS}(\xi, u) = \int_0^\infty E\big(e^{i\xi N + iuXS}|S = s\big)f_S(s)ds$$

or equivalently

$$\varphi_{N,XS}(\xi, u) = \int_0^\infty E\big(e^{i\xi N + iusX}|S = s\big)f_S(s)ds. \tag{2.11.9}$$

From (2.11.9) and the independence of the random variables N, X, S. It follows that

$$\varphi_{N,XS}(\xi, u) = \int_0^\infty E\big(e^{i\xi N + iusX}\big)f_S(s)ds. \tag{2.11.10}$$

Since the independence of the random variables N, X, S implies the independence of the random variables N, X then the random variables $e^{i\xi N}$, e^{iusX} are also independent. Hence (2.11.10) has the form

$$\varphi_{N,XS}(\xi, u) = \int_0^\infty E\big(e^{i\xi N}\big)E\big(e^{iusX}\big)f_S(s)ds$$

or equivalently the form

$$\varphi_{N,XS}(\xi, u) = E\big(e^{i\xi N}\big)\int_0^\infty E\big(e^{iusX}\big)f_S(s)ds. \tag{2.11.11}$$

Since $\varphi_N(\xi) = E(e^{i\xi N})$ and $\varphi_X(us) = E(e^{iusX})$ then (2.11.11) implies that

$$\varphi_{N,XS}(\xi, u) = \varphi_N(\xi) \int_0^\infty \varphi_X(us) f_S(s) ds. \tag{2.11.12}$$

From (2.11.5) it follows that the characteristic function of the random variable XS is

$$\varphi_{XS}(u) = \int_0^\infty \varphi_X(us) f_S(s) ds. \tag{2.11.13}$$

From (2.11.12) and (2.11.13) it follows that (2.11.7) is valid that is $\varphi_{N,XS}(\xi, u) = \varphi_N(\xi)\varphi_{XS}(u)$. Hence the random variables N, XS are independent.

The characteristic function $\varphi_{XS}(u)$ of the random variable XS, the probability function $P(N = n) = p_n$, $n = 0, 1, 2, \ldots$ of the random variable N and the independence of the random variables N, XS permit the evaluation of the characteristic function $\varphi_R(u)$ of the random variable $R = NXS$ in the following way. We have $\varphi_R(u) = E(e^{iuNXS})$ or equivalently

$$\varphi_R(u) = E\big(E\big(e^{iuNXS}|N\big)\big). \tag{2.11.14}$$

From (2.11.14) it follows that

$$\varphi_R(u) = \sum_{n=0}^{\infty} E\big(e^{iuNXS}|N = n\big)P(N = n)$$

or equivalently

$$\varphi_R(u) = \sum_{n=0}^{\infty} E\big(e^{iunXS}|N = n\big)p_n. \tag{2.11.15}$$

Since the random variable N is independent of the random variable XS then (2.11.15) has the form

$$\varphi_R(u) = \sum_{n=0}^{\infty} E\big(e^{iunXS}\big)p_n. \tag{2.11.16}$$

Since $\varphi_{XS}(u) = E(e^{iuXS})$ or equivalently

$$\varphi_{XS}(u) = \int_0^\infty \varphi_X(us) f_S(s) ds$$

then $\varphi_{XS}(nu) = E(e^{inuXS})$ or equivalently

$$\varphi_{XS}(nu) = \int_0^\infty \phi_X(nus)f_S(s)ds. \quad (2.11.17)$$

From (2.11.16) and (2.11.17) it follows that the characteristic function $\varphi_R(u)$ of riskiness $R = NXS$ has the form

$$\varphi_R(u) = \sum_{n=0}^{\infty} p_n \int_0^\infty \varphi_X(nus)f_S(s)ds.$$

Since (2.11.6) has the form

$$\varphi_{XS}(u) = \int_0^\infty \varphi_S(ux)f_X(x)dx$$

then for the characteristic function $\varphi_R(u)$ of riskiness $R = NXS$ is also valid the formula

$$\varphi_R(u) = \sum_{n=0}^{\infty} p_n \int_0^\infty \varphi_S(nux)f_X(x)dx.$$

From a theoretical and a practical point of view it is completely understood that the evaluation of the characteristic function $\varphi_R(u)$ of riskiness $R = NXS$, if the random variables N, X, S are independent, is a very important factor for the probabilistic description, investigation, and solution of problems related with operations of analysis, measurement, evaluation, communication, control, and financing of risks.

The riskiness $R = NXS$ has particular practical interest if risk frequency N follows the Bernoulli distribution. In this case the characteristic function of riskiness has the form

$$\varphi_R(u) = q + p\int_0^\infty \varphi_X(us)f_S(s)ds. \quad (2.11.18)$$

The significance of Bernoulli distribution in the probabilistic consideration of risk frequency makes necessary the evaluation of some special cases of (2.11.18).

We suppose that risk severity X follows the uniform distribution with characteristic function

$$\varphi_X(u) = \frac{e^{iu} - 1}{iu}$$

and risk duration S follows the beta distribution with probability density function $f_S(s) = 2s,\ 0 < s < 1$.

In this case (2.11.18) implies that the characteristic function of riskiness has the form

$$\varphi_R(u) = q + 2p \int_0^1 \frac{e^{ius} - 1}{ius} s ds$$

or equivalently the form

$$\varphi_R(u) = q + 2p \frac{1 + iu - e^{iu}}{u^2}.$$

We suppose that risk severity X follows the exponential distribution with characteristic function

$$\varphi_X(u) = \frac{\mu}{\mu - iu},\ \mu > 0$$

and risk duration S follows the uniform distribution with probability density function $f_S(s) = 1,\ 0 < s < 1$.

The characteristic function of riskiness has the form

$$\varphi_R(u) = q + p \int_0^1 \frac{\mu}{\mu - ius} ds$$

or equivalently the form

$$\varphi_R(u) = q + p \frac{\mu}{iu} \log\left(\frac{\mu}{\mu - iu}\right).$$

We suppose that risk severity X follows the gamma distribution with characteristic function

$$\varphi_X(u) = \left(\frac{\mu}{\mu - iu}\right)^2,\ \mu > 0$$

and risk duration follows S follows the uniform distribution with probability density function $f_S(s) = 1,\ 0 < s < 1$

In this case (2.11.18) implies that the characteristic function of riskiness has the form

$$\varphi_R(u) = q + p\int_0^1 \left(\frac{\mu}{\mu - ius}\right)^2 ds.$$

or equivalently the form

$$\varphi_R(u) = q + p\frac{\mu}{\mu - iu}.$$

We suppose that risk severity X follows the renewal distribution corresponding to the gamma distribution with parameters μ and 2. The characteristic function of the random variable X is

$$\varphi_X(u) = \mu\left[\left(\frac{\mu}{\mu - iu}\right)^2 - 1\right] \Big/ 2iu. \tag{2.11.19}$$

From (2.11.19) it follows that

$$\varphi_X(u) = \frac{1}{2}\left(\frac{\mu}{\mu - iu}\right)^2 + \frac{1}{2}\frac{\mu}{\mu - iu}.$$

Moreover, we suppose that risk duration S follows the uniform distribution with probability density function $f_S(s) = 1,\ 0 < s < 1.$

In this case we have that

$$\varphi_R(u) = q + \frac{p}{2}\int_0^1 \left(\frac{\mu}{\mu - ius}\right)^2 ds + \frac{p}{2}\int_0^1 \frac{\mu}{\mu - ius} ds. \tag{2.11.20}$$

From (2.11.20) it follows that

$$\varphi_R(u) = q + \frac{p}{2}\left(\frac{\mu}{\mu - iu}\right) + \frac{p}{2}\frac{\mu}{iu}\log\left(\frac{\mu}{\mu - iu}\right).$$

The establishment of a sufficient condition for the unimodality at the point 0 of the probability density function $f_R(r)$ of riskiness $R = NXS$ is based on the integral representation of the characteristic function of a probability density function with unique mode at the point 0 and the unimodality at the point 0 of the probability density function of a continuous random variable which is the product of two continuous independent random variables, one of which has a probability density function with unique mode at the point 0.

The existence of a unique mode at the point 0 for the probability density function $f_R(r)$ of riskiness $R = NXS$ substantially facilitates decision making for operations treating a risk of which the frequency is represented by the discrete random variable N with values in the set $\mathbf{N}_0 = \{0, 1, 2, \ldots\}$, the severity is represented by the continuous and positive random variable X and the duration is represented by the continuous and positive random variable S.

The unimodality at the point 0 of the probability density function $f_R(r)$ of riskiness $R = NXS$ can be used for the study of very complex risks related with the evolution of modern organizations. □

Theorem 2.11.2 *We suppose that N is a discrete random variable with values in the set $\mathbf{N}_0 = \{0, 1, 2, \ldots\}$, and probability function $P(N = n) = p_n$, $n = 0, 1, 2, \ldots$, X is a continuous and positive random variable with probability density function $f_X(x)$ and characteristic function $\varphi_X(u)$, and S is a continuous and positive random variable with probability density function $f_S(s)$ and characteristic function $\varphi_S(u)$.*

If the random variables N, X, S are independent and the probability density function $f_X(x)$ is unimodal at the point 0 or the probability density function $f_S(s)$ is unimodal at the point 0 then the probability density function $f_R(r)$ of the random variable $R = NXS$ is unimodal at the point 0.

Proof We suppose that the continuous and positive random variable X has probability density function $f_X(x)$ which is unimodal at the point 0. Since the independence of the random variables N, X, S implies the independence of the random variables X, S then the random variable XS has probability density function which is unimodal at the point 0. Hence the characteristic function

$$\varphi_{XS}(u) = \int_0^\infty \varphi_X(us) f_S(s) ds \tag{2.11.21}$$

of the random variable XS has the form

$$\varphi_{XS}(u) = \int_0^1 \varphi_H(uy) dy \tag{2.11.22}$$

where $\varphi_H(u)$ is the characteristic function of a continuous and positive random variable H.

We consider the sequence of continuous and positive random variables

$$\{nXS, n = 0, 1, 2, \ldots\}. \tag{2.11.23}$$

From (2.11.21) and (2.11.22) it is obvious that the corresponding sequence of characteristic functions of the sequence (2.11.23) is

$$\left\{ \int_0^\infty \varphi_X(nus) f_S(s) ds, n = 0, 1, 2, \ldots \right\} \tag{2.11.24}$$

or equivalently

$$\left\{ \int_0^1 \varphi_H(nuy) dy, n = 0, 1, 2, \ldots \right\}. \tag{2.11.25}$$

From the independence of the random variables N, X, S and Theorem 2.11.1 it follows that the characteristic function of the random variable $R = NXS$ is

$$\varphi_R(u) = \sum_{n=0}^\infty p_n \int_0^\infty \varphi_X(nus) f_S(s) ds. \tag{2.11.26}$$

From (2.11.24), (2.11.25) and (2.11.26) it follows that

$$\varphi_R(u) = \sum_{n=0}^\infty p_n \int_0^1 \varphi_H(nuy) dy$$

or equivalently

$$\varphi_R(u) = \int_0^1 \left(\sum_{n=0}^\infty p_n \varphi_H(nuy) \right) dy. \tag{2.11.27}$$

We consider the sequence of continuous and positive random variables

$$\{nH, n = 0, 1, 2, \ldots\}. \tag{2.11.28}$$

It is obvious that the corresponding sequence of characteristic functions of the sequence (2.11.28) is $\{\varphi_H(nu), n = 0, 1, 2, \ldots\}$.

We consider the function

$$\varphi_V(u) = \sum_{n=0}^\infty p_n \varphi_H(nu)$$

which is a discrete mixture of the characteristic functions of the sequence $\{\varphi_H(nu), n = 0, 1, 2, \ldots\}$ with mixing probability function $p_n, n = 0, 1, 2, \ldots$ which belongs to the random variable N.

Hence the function

$$\varphi_V(u) = \sum_{n=0}^{\infty} p_n \varphi_H(nu) \tag{2.11.29}$$

is the characteristic function of a continuous and positive random variable V.

From (2.11.27) and (2.11.29) it follows that

$$\varphi_R(u) = \int_0^1 \varphi_V(uy)dy. \tag{2.11.30}$$

Hence (2.10.8) and (2.11.30) imply that the random variable $R = NXS$ has probability density function $f_R(r)$ which is unimodal at the point 0.

It is obvious that if the assumption of unimodality at the point 0 for the probability density function $f_X(x)$ of the random variable X is replaced by the assumption of unimodality at the point 0 for the probability density function $f_S(s)$ of the random variable S then the random variable $R = NXS$ has probability density function $f_R(r)$ which is also unimodal at the point 0.

The unimodality at the point 0 of the probability density function $f_R(r)$ means that the probability of the event for the riskiness $R = NXS$ to be in an area right to the point 0 is significant.

The independence of the random variables N, X, S implies that the mean value of the riskiness $R = NXS$ is $E(R) = E(N)E(X)E(S)$. In the case of independence of the random variables N, X, S the evaluation of the mean value of riskiness $R = NXS$ is based on the characteristic function

$$\varphi_R(u) = \sum_{n=0}^{\infty} p_n \int_0^{\infty} \varphi_X(nus) f_S(s) ds \tag{2.11.31}$$

of riskiness $R = NXS$.

From (2.11.31) we get that

$$\varphi'_R(u) = \sum_{n=0}^{\infty} np_n \int_0^{\infty} \varphi'_X(nus) s f_S(s) ds. \tag{2.11.32}$$

Moreover, from (2.11.32) it follows that

$$\varphi'_R(0) = \sum_{n=0}^{\infty} np_n\varphi'_X(0) \int_0^{\infty} sf_S(s)ds$$

or equivalently $E(R) = E(N)E(X)E(S)$. If the probabilistic information for the independent random variables N, X, S are provided by the mean values $E(N)$, $E(X)$, $E(S)$ then the mean value of riskiness $E(R) = E(N)E(X)E(S)$ is a very useful analytical tool for the development and application of risk classification operations. □

2.12 Total Risk Severity and Asset Liquidation

Let N be a discrete random variable with values in the set $\mathbf{N}_0$ and probability generating function $P_N(z)$. Let $\{X_n, n = 1, 2, \ldots\}$ be a sequence of continuous, positive, independent, and identically distributed random variables. The random variables of the sequence $\{X_n, n = 1, 2, \ldots\}$ are equally distributed with the random variable X having characteristic function $\varphi_X(u)$.

We set $T = X_1 + X_2 + \cdots + X_N$.

Let $\{C_n, n = 1, 2, \ldots\}$ be a sequence of continuous, positive, independent, and identically distributed random variables. The random variables of the sequence $\{C_n, n = 1, 2, \ldots\}$ are equally distributed with the random variable C having characteristic function $\varphi_C(\xi)$.

We set $L = C_1 + C_2 + \cdots + C_N$.

We consider the vector (T, L).

The purpose of the present section is the establishment of properties and applications in risk management of the above vector.

The following result establishes sufficient conditions for the evaluation of the characteristic function $\varphi_T(u, \xi)$ of the vector (T, L).

Theorem 2.12.1 *Let $\{X_n, n = 1, 2, \ldots\}$ be a sequence of continuous, positive, independent, and identically distributed random variables. The random variables of the sequence $\{X_n, n = 1, 2, \ldots\}$ are equally distributed with the random variable X having characteristic function*

$$\varphi_X(u). \tag{2.12.1}$$

We consider the discrete random variable N with values in the set $\mathbf{N}_0 = \{0, 1, 2, \ldots\}$ and probability generating function

$$P_N(z) \tag{2.12.2}$$

and we set $T = X_1 + X_2 + \cdots + X_N$.

Let $\{C_n, n = 1, 2, \ldots\}$ *be a sequence of continuous, positive, independent, and identically distributed random variables. The random variables of the sequence* $\{C_n, n = 1, 2, \ldots\}$ *are equally distributed with the random variable* C *having characteristic function*

$$\varphi_C(\xi). \tag{2.12.3}$$

We set $L = C_1 + C_2 + \cdots + C_N$. If $\{X_n, n = 1, 2, \ldots\}$, N *and* $\{C_n, n = 1, 2, \ldots\}$ *are independent then the characteristic function of the vector* (T, L) *is* $\varphi_{T,L}(u, \xi) = P_N(\varphi_X(u)\varphi_C(\xi))$.

Proof We have $\varphi_{T,L}(u, \xi) = E(e^{iuT+i\xi L})$ or equivalently we have

$$\varphi_{T,L}(u, \xi) = E(E(e^{iuT+i\xi L}|N)). \tag{2.12.4}$$

From (2.12.4) it follows that

$$\varphi_{T,L}(u, \xi) = \sum_{n=0}^{\infty} E(e^{iuT+i\xi L}|N = n)P(N = n). \tag{2.12.5}$$

It is easily seen that (2.12.5) implies that

$$\varphi_{T,L}(u, \xi) = \sum_{n=0}^{\infty} E\Big(e^{iu(X_1+\cdots+X_N)+i\xi(C_1+\cdots+C_N)}|N = n\Big)P(N = n). \tag{2.12.6}$$

From (2.12.6) we get that

$$\varphi_{T,L}(u, \xi) = \sum_{n=0}^{\infty} E\Big(e^{iu(X_1+\cdots+X_n)+i\xi(C_1+\cdots+C_n)}|N = n\Big)P(N = n). \tag{2.12.7}$$

Moreover (2.12.7) implies that

$$\varphi_{T,L}(u, \xi) = \sum_{n=0}^{\infty} E(e^{iuX_1+\cdots+iuX_n+i\xi C_1+\cdots+i\xi C_n}|N = n)P(N = n). \tag{2.12.8}$$

From (2.12.8) it follows that

$$\varphi_{T,L}(u, \xi) = \sum_{n=0}^{\infty} E(e^{iuX_1}\ldots e^{iuX_n}e^{iuC_1}\ldots e^{iuC_n}|N = n)P(N = n). \tag{2.12.9}$$

From the assumption that $\{X_n, n = 1, 2, \ldots\}$, N and $\{C_n, n = 1, 2, \ldots\}$ are independent it follows the independence of the random variables $X_1, \ldots, X_n$, N, $C_1, \ldots, C_n$.

The independence of the above random variables implies the independence of the random variables $e^{iuX_1}, \ldots, e^{iuX_n}, \ldots, N, e^{i\xi C_1}, \ldots, e^{i\xi C_n}$.

Hence (2.12.9) has the form

$$\varphi_{T,L}(u,\xi) = \sum_{n=0}^{\infty} E(e^{iuX_1} \ldots e^{iuX_n} e^{i\xi C_1} \ldots e^{i\xi C_n}) P(N=n). \tag{2.12.10}$$

Moreover, the independence of the random variables $e^{iuX_1}, \ldots, e^{iuX_n}, \ldots, N, e^{i\xi C_1}, \ldots, e^{i\xi C_n}$ implies the independence of the random variables $e^{iuX_1}, \ldots, e^{iuX_n}, \ldots, e^{i\xi C_1}, \ldots, e^{i\xi C_n}$.

Hence (2.12.10) has the form

$$\varphi_{T,L}(u,\xi) = \sum_{n=0}^{\infty} E(e^{iuX_1}) \ldots E(e^{iuX_n}) E(e^{i\xi C_1}) \ldots E(e^{i\xi C_n}) P(N=n). \tag{2.12.11}$$

Since the random variables of the sequence $\{X_n, n = 1, 2, \ldots\}$ are equally distributed with the random variable X and the random variables of the sequence $\{C_n, n = 1, 2, \ldots\}$ are equally distributed with the random variable C then (2.12.1), (2.12.3) and (2.12.11) imply that

$$\varphi_{T,L}(u,\xi) = \sum_{n=0}^{\infty} \varphi_X^n(u) \varphi_C^n(\xi) P(N=n). \tag{2.12.12}$$

From (2.12.2) and (2.12.12) it follows that the characteristic function of the vector $(T, L)$is

$$\varphi_{T,L}(u,\xi) = P_N(\varphi_X(u)\varphi_C(\xi)).$$

An interpretation of the vector (T, L), where $T = X_1 + X_2 + \cdots + X_N$ and $L = C_1 + C_2 + \cdots + C_N$, in risk management is the following.

We consider a firm under conditions of risk and asset liquidation in a given time interval. We suppose that the random variable N denotes the frequency of risk in that time interval. The random variable X_n denotes the economic loss due to the nth occurrence of risk. Hence the random variable $T = X_1 + X_2 + \cdots + X_N$ denotes the risk severity in the given time interval. We suppose that the random variable C_n denotes the income of the firm from asset liquidation at the time of the nth occurrence of the risk. Hence the random variable $L = C_1 + C_2 + \cdots + C_N$ denotes the total income of the firm from asset liquidation in the given time interval. In this case the vector (T, L) constitutes a strong analytical tool for investigating the evolution of the firm under conditions of risk and asset liquidation in a given time interval. The independence for $\{X_n, n = 1, 2, \ldots\}$ N and $\{C_n, n = 1, 2, \ldots\}$ is a sufficient condition for evaluating the characteristic function $\varphi_{T,L}(u,\xi) = P_N(\varphi_X(u)\varphi_C(\xi))$ of the vector (T, L).

In this case the applicability of the above vector is substantially extended. □

2.13 Total Risk Severity and Total Income

Let $\{X_n, n = 1, 2, \ldots\}$ be a sequence of continuous, positive, independent, and identically distributed random variables. The random variables of the sequence $\{X_n, n = 1, 2, \ldots\}$ are equally distributed with the random variable X having characteristic function $\varphi_X(u)$.

We consider the discrete random variable N with values in the set $\mathbf{N}_0 = \{0, 1, 2, \ldots\}$ and probability generating function $P_N(z)$.

We set $T = X_1 + X_2 + \cdots + X_N$.

Let $\{C_s, s = 1, 2, \ldots\}$ be a sequence of continuous, positive, independent, and identically distributed random variables. The random variables of the sequence $\{C_s, s = 1, 2, \ldots\}$ are equally distributed with the random variable C having characteristic function $\varphi_C(\xi)$.

We consider the discrete random variable S with values in the set $\mathbf{N}_0 = \{0, 1, 2, \ldots\}$ and probability generating function $P_S(z)$ and we set $L = C_1 + C_2 + \cdots + C_S$.

We consider the vector (T, L).

The purpose of the present section is the establishment of properties and applications in risk management of the above vector.

The following theorem establishes sufficient conditions for the evaluation of the characteristic function of the vector (T, L).

Theorem 2.13.1 *Let $\{X_n, n = 1, 2, \ldots\}$ be a sequence of continuous, positive, independent, and identically distributed random variables. The random variables of the sequence $\{X_n, n = 1, 2, \ldots\}$ are equally distributed with the random variable X having characteristic function*

$$\varphi_X(u). \tag{2.13.1}$$

We consider the discrete random variable N with values in the set $\mathbf{N}_0 = \{0, 1, 2, \ldots\}$ and probability generating function

$$P_N(z) \tag{2.13.2}$$

and we set $T = X_1 + X_2 + \cdots + X_N$.

Let $\{C_s, s = 1, 2, \ldots\}$ be a sequence of continuous, positive, independent, and identically distributed random variables. The random variables of the sequence $\{C_s, s = 1, 2, \ldots\}$ are equally distributed with the random variable C having characteristic function

$$\varphi_C(\xi). \tag{2.13.3}$$

We consider the discrete random variable S with values in the set $\mathbf{N}_0 = \{0, 1, 2, \ldots\}$ and probability generating function

$$P_S(z) \tag{2.13.4}$$

and we set $L = C_1 + C_2 + \cdots + C_S$.

If $\{X_n, n = 1, 2, \ldots\}$, N, $\{C_s, s = 1, 2, \ldots\}$ and S are independent then the random variables $T = X_1 + X_2 + \cdots + X_N$ and $L = C_1 + C_2 + \cdots + C_S$ are independent and $\varphi_{T,L}(u, \xi) = P_N(\varphi_X(u)P(\varphi_C(\xi)))$ is the characteristic function of the vector (T, L).

Proof The independence of $\{X_n, n = 1, 2, \ldots\}$, N, $\{C_s, s = 1, 2, \ldots\}$ and S implies the independence of $\{X_n, n = 1, 2, \ldots\}$ and N, and the independence of $\{C_s, s = 1, 2, \ldots\}$ and S.

Hence (2.13.1) and (2.13.2) imply that the characteristic function of the random variable $T = X_1 + X_2 + \cdots + X_N$ is

$$\varphi_T(u) = P_N(\varphi_X(u)) \tag{2.13.5}$$

and (2.13.3), (2.13.4) imply that the characteristic function of the random variable $L = C_1 + C_2 + \cdots + C_S$ is

$$\varphi_L(\xi) = P_S(\varphi_C(\xi)). \tag{2.13.6}$$

Let $\varphi_{T,L}(u, \xi)$ be the characteristic function of the vector (T, L).

The establishment of the independence of the random variables $T = X_1 + X_2 + \cdots + X_N$ and $L = C_1 + C_2 + \cdots + C_S$ requires the establishment of the relationship $\varphi_{T,L}(u, \xi) = \varphi_T(u)\varphi_L(\xi)$.

We have $\varphi_{T,L}(u, \xi) = E\big(e^{iuT+i\xi L}\big)$ or equivalently

$$\varphi_{T,L}(u, \xi) = E\big(E\big(e^{iuT+i\xi L}|N, S\big)\big). \tag{2.13.7}$$

From (2.13.7) it follows that

$$\varphi_{T,L}(u, \xi) = \sum_{n=0}^{\infty}\sum_{s=0}^{\infty} E\big(e^{iuT+i\xi L}|N = n, S = s\big)P(N = n, S = s)$$

or equivalently

$$\varphi_{T,L}(u, \xi) = \sum_{n=0}^{\infty}\sum_{s=0}^{\infty} E\Big(e^{iu(X_1+\cdots+X_N)+i\xi(C_1+\cdots+C_S)}|N = n, S = s\Big)P(N = n, S = s). \tag{2.13.8}$$

From (2.13.8) it follows that

$$\varphi_{T,L}(u,\xi) = \sum_{n=0}^{\infty}\sum_{s=0}^{\infty} E\left(e^{iu(X_1+\cdots+X_n)+i\xi(C_1+\cdots+C_s)}|N=n, S=s\right)P(N=n, S=s). \tag{2.13.9}$$

Moreover, (2.13.9) implies that

$$\varphi_{T,L}(u,\xi) = \sum_{n=0}^{\infty}\sum_{s=0}^{\infty} E(e^{iuX_1}\ldots e^{iuX_n}, e^{i\xi C_1}\ldots e^{i\xi C_s}|N=n, S=s)P(N=n, S=s). \tag{2.13.10}$$

The independence of $\{X_n, n=1,2,\ldots\}$, N, $\{C_s, s=1,2,\ldots\}$ and S implies the independence of the random variables $X_1,\ldots,X_n$, N, $C_1,\ldots,C_s$, S.

The independence of the above random variables implies the independence of the random variables $e^{iuX_1},\ldots,e^{iuX_n}$, N, $e^{i\xi C_1},\ldots,e^{i\xi C_s}$, S.

Hence (2.13.10) has the form

$$\varphi_{T,L}(u,\xi) = \sum_{n=0}^{\infty}\sum_{s=0}^{\infty} E(e^{iuX_1}\ldots e^{iuX_n}e^{i\xi C_1}\ldots e^{i\xi C_s})P(N=n, S=s). \tag{2.13.11}$$

The independence of the random variables $e^{iuX_1},\ldots,e^{iuX_n}$, N, $e^{i\xi C_1},\ldots,e^{i\xi C_s}$, S implies the independence of the random variables $e^{iuX_1},\ldots,e^{iuX_n}$, $e^{i\xi C_1},\ldots,e^{i\xi C_s}$, and the independence of the random variables N, S.

Hence (2.13.11) has the form

$$\varphi_{T,L}(u,\xi) = \sum_{n=0}^{\infty}\sum_{s=0}^{\infty} E(e^{iuX_1})\ldots E(e^{iuX_n})E(e^{i\xi C_1})\ldots E(e^{iuC_s})P(N=n)P(S=s)$$

or equivalently the form

$$\varphi_{T,L}(u,\xi) = \sum_{n=0}^{\infty} E(e^{iuX_1})\ldots E(e^{iuX_n})P(N=n)\sum_{s=0}^{\infty} E(e^{i\xi C_1})\ldots E(e^{i\xi C_s})P(S=s). \tag{2.13.12}$$

Since the random variables of the sequence $\{X_n, n=1,2,\ldots\}$ are equally distributed with the random variable X having characteristic function $\varphi_X(u)$ and the random variables of the sequence $\{C_s, s=1,2,\ldots\}$ are equally distributed with the random variable C having characteristic function $\varphi_C(\xi)$ then (2.13.12) has the form

$$\varphi_{T,L}(u,\xi) = \sum_{n=0}^{\infty} \varphi_X^n(u)P(N=n)\sum_{s=0}^{\infty} \varphi_C^s(\xi)P(S=s). \tag{2.13.13}$$

From (2.13.1), (2.13.2), (2.13.3),(2.13.4) and (2.13.13) it follows that

$$\varphi_{T,L}(u,\xi) = P_N(\varphi_X(u))P_S(\varphi_C(\xi)). \tag{2.13.14}$$

Moreover (2.13.5), (2.13.6) and (2.13.14) imply that $\varphi_{T,L}(u,\xi) = \varphi_T(u)\varphi_L(\xi)$.

Hence the random variables $T = X_1 + X_2 + \cdots + X_N$ and $L = C_1 + C_2 + \cdots + C_S$ are independent and the characteristic function of the vector (T,L) is $\varphi_{T,L}(u,\xi) = P_N(\varphi_X(u))P_S(\varphi_C(\xi))$.

The interpretation of the vector (T,L) as the concept of a quantitative component of risk is the following.

We consider a firm under the occurrences of a risk and the creation of incomes in a given time interval. We suppose that the random variable N denotes the frequency of risk in that time interval and the random variable X_n denotes the economic loss due to the nth occurrence of risk. Hence the random variable $T = X_1 + X_2 + \cdots + X_N$ denotes the total risk severity in the given time interval. We suppose that the random variable S denotes the number of incomes created by the firm in the same time interval and the random variable C_s denotes the size of the sth income created by the production activities of the firm. Hence the random variable $L = C_1 + C_2 + \cdots + C_S$ denotes the total income created by the firm in that time interval. In this case, the vector (T,L), where $T = X_1 + X_2 + \cdots + X_N$ and $L = C_1 + C_2 + \cdots + C_S$ constitutes a strong analytical tool for investigating the behavior of the firm under the occurrences of a risk and the creation of incomes in a given time interval. The independence of $\{X_n, n = 1,2,\ldots\}$, N, $\{C_s, s = 1,2,\ldots\}$ and S is a sufficient condition for evaluating the characteristic function $\varphi_{T,L}(u,\xi) = P_N(\varphi_X(u))P_S(\varphi_C(\xi))$ of the vector (T,L).

In this case the applicability of the above vector is substantially extended. □

2.14 Recovery Time of a Partially Damaged System and Release Time of a Backup System

Let N be a discrete random variable with values in the set $\mathbf{N} = \{1,2,\ldots\}$ and probability generating function $P_N(z)$.

We suppose that $\{X_n, n = 1,2,\ldots\}$ is a sequence of continuous, positive, independent, and identically distributed random variables. The random variables of the sequence are equally distributed with the random variable X having distribution function $F_X(x)$.

We set $T = \max(X_1, X_2, \ldots, X_N)$.

Let $\{C_n, n = 1,2,\ldots\}$ be a sequence of continuous, positive, independent, and identically distributed random variables. The random variables of the sequence are equally distributed with the random variable C having distribution function $F_C(c)$.

We set $L = \max(C_1, C_2, \ldots, C_N)$.

We consider the vector (T,L).

The purpose of the present section is the establishment of properties and applications in risk management of the vector (T, L).

The following result establishes sufficient conditions for evaluating the distribution function $F_{T,L}(t,\ell)$ of the vector (T, L).

Theorem 2.14.1 *Let* N *be a discrete random variable with values in the set* $\mathbf{N} = \{1, 2, \ldots\}$ *and probability generating function*

$$P_N(z). \tag{2.14.1}$$

We suppose that $\{X_n, n = 1, 2, \ldots\}$ *be a sequence of continuous, positive, independent, and identically distributed random variables. Moreover, we suppose that the random variables of the sequence are equally distributed with the random variable* X *having distribution function*

$$F_X(x) \tag{2.14.2}$$

and we set $T = \max(X_1, X_2, \ldots, X_N)$.

Let $\{C_n, n = 1, 2, \ldots\}$ *be a sequence of continuous, positive, independent, and identically distributed random variables. Moreover, we suppose that the random variables of the sequence are equally distributed with the random variable* C *having distribution function*

$$F_C(c) \tag{2.14.3}$$

and we set $L = \max(C_1, C_2, \ldots, C_N)$.

We consider the vector (T, L). *If* N, $\{X_n, n = 1, 2, \ldots\}$ *and* $\{C_n, n = 1, 2, \ldots\}$ *are independent then the distribution function* $F_{T,L}(t,\ell)$ *of the vector* (T, L) *is* $F_{T,L}(t,\ell) = P_N(F_X(t)F_C(\ell))$.

Proof We have $F_{T,L}(t,\ell) = P(T \le t, L \le \ell)$ or equivalently

$$F_{T,L}(t,\ell) = \sum_{n=1}^{\infty} P(T \le t, L \le \ell | N = n)P(N = n). \tag{2.14.4}$$

From (2.14.4) it follows that

$$F_{T,L}(t,\ell) = \sum_{n=1}^{\infty} P[\max(X_1, X_2, \ldots, X_N) \le t, \max(C_1, C_2, \ldots, C_N) \le \ell | N = n]P(N = n). \tag{2.14.5}$$

Hence (2.14.5) implies that

$$F_{T,L}(t,\ell) = \sum_{n=1}^{\infty} P[\max(X_1, \ldots, X_n) \le t, \max(C_1, \ldots, C_n) \le \ell | N = n]P(N = n). \tag{2.14.6}$$

From (2.14.6) it follows that

$$F_{T,L}(t,\ell) = \sum_{n=1}^{\infty} P(X_1 \le t,\ldots,X_n \le t, C_1 \le \ell,\ldots,C_n \le \ell | N = n)P(N = n). \tag{2.14.7}$$

The independence of N, $\{X_n, n = 1,2,\ldots\}$ and $\{C_n, n = 1,2,\ldots\}$ implies the independence of the random variables $N,\ X_1,\ldots,X_n,\ C_1,\ldots,C_n$.

Hence (2.14.7) has the form

$$F_{T,L}(t,\ell) = \sum_{n=1}^{\infty} P(X_1 \le t,\ldots,X_n \le t, C_1 \le \ell,\ldots,C_n \le \ell)P(N = n). \tag{2.14.8}$$

From the independence of the random variables $N,\ X_1,\ldots,X_n,\ C_1,\ldots,C_n$. It follows the independence of the random variables $X_1,\ldots,X_n,\ C_1,\ldots,C_n$.

The independence of the above random variables implies that (2.14.8) has the form

$$F_{T,L}(t,\ell) = \sum_{n=1}^{\infty} P(X_1 \le t)\ldots P(X_n \le t)P(C_1 \le \ell)\ldots P(C_n \le \ell)P(N = n). \tag{2.14.9}$$

Since the random variables of the sequence $\{X_n, n = 1,2,\ldots\}$ are equally distributed with the random variable X and the random variables of the sequence $\{C_n, n = 1,2,\ldots\}$ are equally distributed with the random variable C then (2.14.2), (2.14.3) and (2.14.9) imply that

$$F_{T,L}(t,\ell) = \sum_{n=1}^{\infty} F_X^n(t)F_C^n(\ell)P(N = n). \tag{2.14.10}$$

From (2.14.1) and (2.14.10) it follows that the distribution function $F_{T,L}(t,\ell)$ of the vector (T,L) has the form $F_{T,L}(t,\ell) = P_N(F_X(t)F_C(\ell))$.

An interpretation of the vector (T,L), where $T = \max(X_1,X_2,\ldots,X_N)$ and $L = \max(C_1,C_2,\ldots,C_N)$, in risk management is the following.

We suppose that the occurrence of a risk, at the time point 0, interrupts N operations of a system. This system is called system I. The random variable X_n denotes the time required for the recovery of the nth interrupted operation of system I. Hence the random variable $T = \max(X_1,X_2,\ldots,X_N)$ denotes the time required for the recovery of system I. We suppose that the N operations of system I, which are interrupted by the occurrence of risk at the time point 0, are undertaken by another system which is called system II. The random variable C_n denotes the available time of system II for undertaking the nth interrupted operation of system II. Hence the random variable

$L = \max(C_1, C_2, \ldots, C_N)$ denotes the release time of system II from the undertaking of the N interrupted operations of system II. In this case the vector (T, L) is particularly useful for investigating the behavior of system I.

The independence of N, $\{X_n, n = 1, 2, \ldots\}$ and $\{C_n, n = 1, 2, \ldots\}$ permits the evaluation of the distribution function $F_{T,L}(t, \ell) = P_N(F_X(t)F_C(\ell))$ of the vector (T, L).

In this case the the practical applicability of the above vector in risk management is substantially extended.

The interpretation of the random variable $L = \max(C_1, C_2, \ldots, C_N)$ as the release time of system II from the undertaking of the N interrupted operations of system I means that the consideration of system II and the corresponding vector (T, L) facilitates the development and implementation of risk treatment operations. □

2.15 Vector of Recovery Times of Two Partially Damaged Systems

Let $\{X_n, n = 1, 2, \ldots\}$ be a sequence of continuous, positive, independent, and identically distributed random variables. The random variables of the sequence $\{X_n, n = 1, 2, \ldots\}$ are equally distributed with the random variable X having distribution function $F_X(x)$.

We consider the discrete random variable N with values in the set $\mathbf{N} = \{1, 2, \ldots\}$ and probability generating function $P_N(z)$, and we set $T = \max(X_1, X_2, \ldots, X_N)$.

Let $\{C_s, s = 1, 2, \ldots\}$ be a sequence of continuous, positive, independent, and identically distributed random variables. The random variables of the sequence $\{C_s, s = 1, 2, \ldots\}$ are equally distributed with the random variable C having distribution function $F_C(c)$.

We consider the discrete random variable S with values in the set $\mathbf{N} = \{1, 2, \ldots\}$ and probability generating function $P_S(z)$, and we set $L = \max(C_1, C_2, \ldots, C_S)$.

We consider the vector (T, L).

The purpose of the present section is the establishment properties and applications in risk management of the above vector.

The following result establishes sufficient conditions for evaluating the distribution function of the vector (T, L)

Theorem 2.15.1 *Let $\{X_n, n = 1, 2, \ldots\}$ be a sequence of continuous, positive, independent, and identically distributed random variable. The random variables of the sequence $\{X_n, n = 1, 2, \ldots\}$ are equally distributed with the random variable X having distribution function*

$$F_X(x). \tag{2.15.1}$$

We consider the discrete random variable N with values in the set $\mathbf{N} = \{1, 2, \ldots\}$ *and probability generating function*

$$P_N(z) \tag{2.15.2}$$

and we set $T = \max(X_1, X_2, \ldots, X_N)$.

Let $\{C_s, s = 1, 2, \ldots\}$ *be a sequence of continuous, positive, independent, and identically distributed random variables. The sequence of the random variables* $\{C_s, s = 1, 2, \ldots\}$ *are equally distributed with the random variable C having distribution function*

$$F_C(c). \tag{2.15.3}$$

We consider the discrete random variable S with values in the set $\mathbf{N} = \{1, 2, \ldots\}$ *and probability generating function*

$$P_S(z) \tag{2.15.4}$$

and we set $L = \max(C_1, C_2, \ldots, C_S)$.

If $\{X_n, n = 1, 2, \ldots\}$, N, $\{C_s, s = 1, 2, \ldots\}$ *and S are independent then* $T = \max(X_1, X_2, \ldots, X_N)$ *and* $L = \max(C_1, C_2, \ldots, C_S)$ *are independent and* $F_{T,L}(t, \ell) = P_N(F_X(t))P_S(F_C(\ell))$ *is the distribution function of the vector* (T, L).

Proof The assumption that $\{X_n, n = 1, 2, \ldots\}$, N, $\{C_s, s = 1, 2, \ldots\}$ and S are independent implies that $\{X_n, n = 1, 2, \ldots\}$ and N are independent and also that $\{C_s, s = 1, 2, \ldots\}$ and S are independent. Hence (2.15.1) and (2.15.2) imply that the distribution function of the random variable $T = \max(X_1, X_2, \ldots, X_N)$ is

$$F_T(t) = P_N(F_X(t)) \tag{2.15.5}$$

and (2.15.3), (2.15.4) imlply that the distribution function of the random variable $L = \max(C_1, C_2, \ldots, C_S)$ is

$$F_L(\ell) = P_S(F_C(\ell)). \tag{2.15.6}$$

Let $F_{T,L}(t, \ell)$ be the distribution function of the vector (T, L).

The proof of the independence of the random variables $T = \max(X_1, X_2, \ldots, X_N)$ and $L = \max(C_1, C_2, \ldots, C_S)$ requires the establishment of the relationship $F_{T,L}(t, \ell) = F_T(t)F_L(\ell)$. Since

$$F_{T,L}(t, \ell) = P(T \leq t, L \leq \ell). \tag{2.15.7}$$

We get that

$$F_{T.L}(t, \ell) = \sum_{n=1}^{\infty} \sum_{s=1}^{\infty} P(T \leq t, L \leq \ell | N = n, S = s) P(N = n, S = s). \tag{2.15.8}$$

From (2.15.8) it follows that

$$F_{T,L}(t,\ell) = \sum_{n=1}^{\infty}\sum_{s=1}^{\infty} P(\max(X_1,\ldots,X_N) \le t, \max(C_1,\ldots,C_S) \le \ell | N=n, S=s) P(N=n, S=s)$$

or equivalently

$$F_{T,L}(t,\ell) = \sum_{n=1}^{\infty}\sum_{s=1}^{\infty} P(\max(X_1,\ldots,X_n) \le t, \max(C_1,\ldots,C_s) \le \ell | N=n, S=s) P(N=n, S=s). \tag{2.15.9}$$

From (2.15.9) it follows that

$$F_{T,L}(t,\ell) = \sum_{n=1}^{\infty}\sum_{s=1}^{\infty} P(X_1 \le t,\ldots,X_n \le t, C_1 \le \ell,\ldots,C_s \le \ell | N=n, S=s) P(N=n, S=s). \tag{2.15.10}$$

From the fact that $\{X_n, n=1,2,\ldots\}$, N, $\{C_s, s=1,2,\ldots\}$ and S are independent it follows the independence of the random variables $X_1,\ldots,X_n$, N, $C_1,\ldots,C_s$, S.
Hence (2.15.10) has the form

$$F_{T,L}(t,\ell) = \sum_{n=1}^{\infty}\sum_{s=1}^{\infty} P(X_1 \le t,\ldots,X_n \le t, C_1 \le \ell,\ldots,C_s \le \ell) P(N=n, S=s). \tag{2.15.11}$$

The independence of the random variables $X_1,\ldots,X_n$, N, $C_1,\ldots,C_s$, S implies the independence of the random variables $X_1,\ldots,X_n$, $C_1,\ldots,C_s$ and the independence of the random variables N, S.
Hence (2.15.11) has the form

$$F_{T,L}(t,\ell) = \sum_{n=1}^{\infty}\sum_{s=1}^{\infty} P(X_1 \le t)\ldots P(X_n \le t) P(C_1 \le \ell)\ldots P(C_s \le \ell) P(N=n) P(S=s). \tag{2.15.12}$$

Since the random variables of the sequence $\{X_n, n=1,2,\ldots\}$ are equally distributed with the random variable X having distribution function $F_X(x)$ and the random variables of the sequence $\{C_s, s = 1, 2, \ldots\}$ are equally distributed with the random variable C having distribution function $F_C(c)$ then (2.15.12) has the form

$$F_{T,L}(t,\ell) = \sum_{n=1}^{\infty} P(X_1 \leq t) \ldots P(X_n \leq t) P(N = n) \sum_{s=1}^{\infty} P(C_1 \leq \ell) \ldots P(C_s \leq \ell) P(S = s)$$

or equivalently the form

$$P(T \leq t, L \leq \ell) = \sum_{n=1}^{\infty} F_X^n(t) P(N = n) \sum_{s=1}^{\infty} F_C^s(\ell) P(S = s). \tag{2.15.13}$$

From (2.15.13) it follows that

$$F_{T,L}(t,\ell) = P_N(F_X(t)) P_S(F_C(\ell)). \tag{2.15.14}$$

Moreover (2.15.5), (2.15.6), (2.15.7) and (2.15.14) imply that $F_{T,L}(t,\ell) = F_T(t) F_L(\ell)$.

Hence the random variables $T = \max(X_1, X_2, \ldots, X_N)$ and $L = \max(C_1, C_2, \ldots, C_S)$ are independent and the distribution function of the random vector (T, L) is $F_{T,L}(t,\ell) = P_N(F_X(t)) P_S(F_C(\ell))$.

An interpretation of the vector (T, L) where $T = \max(X_1, X_2, \ldots, X_N)$ and $L = \max(C_1, C_2, \ldots, C_S)$ in risk management is the following.

We consider two systems. The occurrence of a risk, at the time point 0, interrupts N operations of the first system and S operations of the second system. The random variable X_n denotes the time required for the recovery of the nth interrupted operation of the first system. Hence the random variable $T = \max(X_1, X_2, \ldots, X_N)$ denotes the time required for the recovery of the first system. The random variable C_s denotes the time required for the recovery of the sth interrupted operation of the second system. Hence the random variable $L = \max(C_1, C_2, \ldots, C_S)$ denotes the time required for the recovery of the second system. In this case the vector (T, L) is particularly useful for investigating the evolution of the pair of two systems. The independence of $\{X_n, n = 1, 2, \ldots\}$, N, $\{C_s, s = 1, 2, \ldots\}$, and S permits the evaluation of the distribution function $F_{T,L}(t,\ell) = P_N(F_X(t)) P_S(F_C(\ell))$ of the vector (T, L).

In this case the practical applicability in risk management of the above vector is substantially extended.

It is obvious that the results of the present section for the vector (T, L) can be extended for vectors of many dimensions. □

2.16 Recovery Time of a Partially Damaged System Under a Random Number of Competing Risks

Let $\{X_n, n = 1, 2, \ldots\}$ be a sequence of continuous, positive, independent, and identically distributed random variables. The random variables of the sequence $\{X_n, n = 1, 2, \ldots\}$ are equally distributed with the random variable X having distribution function $F_X(x)$.

We consider the discrete random variable N with values in the set $\mathbf{N} = \{1, 2, \ldots\}$ and probability generating function $P_N(z)$, and set $T = \max(X_1, X_2, \ldots, X_N)$.

Let $\{C_s, s = 1, 2, \ldots\}$ be a sequence of continuous, positive, independent, and identically distributed random variables. The random variables of the sequence $\{C_s, s = 1, 2, \ldots\}$ are equally distributed with the random variable C having distribution function $F_C(c)$.

We consider the discrete random variable S with values in the set $\mathbf{N} = \{1, 2, \ldots\}$ and probability generating function $P_S(z)$, and set $L = \min(C_1, C_2, \ldots, C_S)$.

We consider the vector (T, L).

The purpose of the present section is the establishment of properties and applications in risk management of the above vector.

The following result establishes sufficient conditions for evaluating the distribution function of the vector (T, L).

Theorem 2.16.1 *Let $\{X_n, n = 1, 2, \ldots\}$ be a sequence of continuous, positive, independent, and identically distributed random variables. The random variables of the sequence $\{X_n, n = 1, 2, \ldots\}$ are equally distributed with the random variable X having distribution function*

$$F_X(x). \tag{2.16.1}$$

We consider the discrete random variable N with values in the set $\mathbf{N} = \{1, 2, \ldots\}$ and probability generating function

$$P_N(z) \tag{2.16.2}$$

and we set $T = \max(X_1, X_2, \ldots, X_N)$.

Let $\{C_s, s = 1, 2, \ldots\}$ be a sequence of continuous, positive, independent, and identically distributed random variables. The random variables of the sequence $\{C_s, s = 1, 2, \ldots\}$ are equally distributed with the random variable C having distribution function

$$F_C(c). \tag{2.16.3}$$

We consider the discrete random variable S with values in the set $\mathbf{N} = \{1, 2, \ldots\}$ and probability generating function

$$P_S(z) \tag{2.16.4}$$

and set $L = \min(C_1, C_2, \ldots, C_S)$.

If $\{X_n, n = 1, 2, \ldots\}$, N, $\{C_s, s = 1, 2, \ldots\}$, and S are independent then the random variables $T = \max(X_1, X_2, \ldots, X_N)$ and $L = \min(C_1, C_2, \ldots, C_S)$ are independentvaı and $F_{T,L}(t, \ell) = P_N(F_X(t))(1 - P_S(1 - F_C(\ell)))$ is the distribution function of the vector (T, L).

Proof The independence of $\{X_n, n = 1, 2, \ldots\}, N, \{C_s, s = 1, 2, \ldots\}$ and S implies the independence of $\{X_n, n = 1, 2, \ldots\}$ and N and the independence of $\{C_s, s = 1, 2, \ldots\}$ and S.

Hence (2.16.1) and (2.16.2) imply that the distribution function of the random variable $T = \max(X_1, X_2, \ldots, X_N)$ is

$$F_T(t) = P_N(F_X(t)) \tag{2.16.5}$$

and (2.16.3), (2.16.4) imply that the distribution function of the random variable $L = \min(C_1, C_2, \ldots, C_S)$ is

$$F_L(\ell) = 1 - P_S(1 - F_C(\ell)). \tag{2.16.6}$$

Let $F_{T,L}(t, \ell)$ be the distribution function of the vector (T, L).

The establishment of independence of the random variables $T = \max(X_1, X_2, \ldots, X_N)$ and $L = \min(C_1, C_2, \ldots, C_S)$ requires the establishment of the relationship $F_{T,L}(t, \ell) = F_T(t) F_L(\ell)$.

We have

$$F_{T,L}(t, \ell) = P(T \le t, L \le \ell). \tag{2.16.7}$$

Since $P(T \le t) = P(T \le t, L \le \ell) + P(T \le t, L > \ell)$ then we get

$$P(T \le t, L \le \ell) = P(T \le t) - P(T \le t, L > \ell). \tag{2.16.8}$$

From (2.16.8) it follows that

$$P(T \le t, L \le \ell) = P(T \le t) - \sum_{n=1}^{\infty} \sum_{s=1}^{\infty} P(T \le t, L > \ell | N = n, S = s) P(N = n, S = s) \tag{2.16.9}$$

Moreover (2.16.5), (2.16.7) and (2.16.9) imply that

$$F_{T,L}(t,\ell) = P_N(F_X(t)) \\ -\sum_{n=1}^{\infty}\sum_{s=1}^{\infty} P[\max(X_1, X_2, \ldots, X_N) \le t, \min(C_1, C_2, \ldots, C_S) > \ell | N = n, S = s] P(N = n, S = s). \tag{2.16.10}$$

From (2.16.10) it follows that

$$F_{T,L}(t,\ell) = P_N(F_X(t)) \\ -\sum_{n=1}^{\infty}\sum_{s=1}^{\infty} P[\max(X_1, X_2, \ldots, X_n) \le t, \min(C_1, C_2, \ldots, C_s) > \ell | N = n, S = s] P(N = n, S = s). \tag{2.16.11}$$

From (2.16.11) it follows that

$$F_{T,L}(t,\ell) = P_N(F_X(t)) \\ -\sum_{n=1}^{\infty}\sum_{s=1}^{\infty} P[X_1 \le t, \ldots, X_n \le t, C_1 > \ell, \ldots, C_s > \ell | N = n, S = s] P(N = n, S = s). \tag{2.16.12}$$

The independence of $\{X_n, n = 1, 2, \ldots\}$, N, $\{C_s, s = 1, 2, \ldots\}$ and S implies the independence of the random variables $X_1, \ldots, X_n$, N, $C_1, \ldots, C_s$, S.

Hence (2.16.12) has the form

$$F_{T,L}(t,\ell) = P_N(F_X(t)) \\ -\sum_{n=1}^{\infty}\sum_{s=1}^{\infty} P(X_1 \le t, \ldots X_n \le t, C_1 > \ell, \ldots, C_s > \ell) P(N = n, S = s). \tag{2.16.13}$$

The independence of the random variables $X_1, \ldots, X_n$, N, $C_1, \ldots, C_s$, S implies the independence of the random variables $X_1, \ldots, X_n$, $C_1, \ldots, C_s$ and the independence of the random variables N, S.

Hence (2.16.13) has the form

$$F_{T,L}(t,\ell) = \\ P_N(F_X(t)) - \sum_{n=1}^{\infty}\sum_{s=1}^{\infty} P(X_1 \le t) \ldots P(X_n \le t) P(C_1 > \ell) \ldots P(C_s > \ell) P(N = n) P(S = s) \tag{2.16.14}$$

From (2.16.14) it follows that

$$F_{T,L}(t,\ell) = \\ P_N(F_X(t)) - \sum_{n=1}^{\infty} P(X_1 \le t) \ldots P(X_n \le t) P(N = n) \sum_{s=1}^{\infty} P(C_1 > \ell) \ldots P(C_s > \ell) P(S = s). \tag{2.16.15}$$

Since the random variables of the sequence $\{X_n, n = 1, 2, \ldots\}$ are equally distributed with the random variable X having distribution function $F_X(x)$ and the random variables of the sequence $\{C_s, s = 1, 2, \ldots\}$ are equally distributed with the random variable C having distribution function $F_C(c)$ then (2.16.15) has the form

$$F_{T,L}(t,\ell) = P_N(F_X(t)) - \sum_{n=1}^{\infty} F_X^n(t) P(N = n) \sum_{s=1}^{\infty} [1 - F_C(\ell)]^s P(S = s)$$

or equivalently the form

$$F_{T,L}(t,\ell) = P_N(F_X(t)) - P_N(F_X(t)) P_S(1 - F_C(\ell)). \tag{2.16.16}$$

From (2.16.16) it follows that

$$F_{T,L}(t,\ell) = P_N(F_X(t))[1 - P_S(1 - F_C(\ell))]. \tag{2.16.17}$$

Moreover (2.16.5), (2.16.6) and (2.16.17) imply that $F_{T,L}(t,\ell) = F_T(t)F_L(\ell)$. Hence the random variable $T = \max(X_1, X_2, \ldots, X_N)$ is independent of the random variable $L = \min(C_1, C_2, \ldots, C_S)$ and the distribution function $F_{T,L}(t,\ell)$ of the vector (T, L) is $F_{T,L}(t,\ell) = P_N(F_X(t))[1 - P_S(1 - F_C(\ell))]$..

An interpretation of the vector (T, L), where $T = \max(X_1, X_2, \ldots, X_N)$ and $L = \min(C_1, C_2, \ldots, C_S)$, as a concept of risk management is the following. A risk occurs at the time point 0. The occurrence of risk interrupts N operations of a system. The random variable X_n denotes the time required for the recovery of the nth interrupted operation of the system. Hence the random variable $T = \max(X_1, X_2, \ldots, X_N)$ denotes the time required for the recovery of the system. The random variable $T = \max(X_1, X_2, \ldots, X_N)$ is a fundamental stochastic model for describing and analyzing the recovery process of a system. Moreover S risks threaten the system at the time point 0. The random variable C_s denotes the occurrence time of the sth risk. Hence the random variable $L = \min(C_1, C_2, \ldots, C_S)$ denotes the minimum of the risk occurrence times. The consideration of a system under a random number S of independent competing risks means the use of the random variable $L = \min(C_1, C_2, \ldots, C_S)$ as a fundamental stochastic model for describing and analyzing the evolution of that system. Hence the vector (T, L) can be considered as a strong analytical tool for investigating the behaviour of a system which has experienced the occurrence of a risk and then the system recovers under a random number of independent competing risks. □

2.17 Considering a System Under a Random Number of Competing Risks

Let N be a discrete random variable with values in the set $\mathbf{N} = \{1, 2, \ldots\}$ and probability generating function $P_N(z)$. We suppose that $\{X_n, n = 1, 2, \ldots\}$ is a sequence of continuous, positive, independent, and identically distributed random variables. The random variables of the sequence are equally distributed with the random variable X having distribution function $F_X(x)$ and we set $T = \min(X_1, X_2, \ldots, X_N)$.

Let $\{C_n, n = 1, 2, \ldots\}$ be a sequence of continuous, positive, independent, and identically distributed random variables. The random variables of the sequence are equally distributed with the random variable C having distribution function $F_C(c)$ and we set $L = \min(C_1, C_2, \ldots, C_N)$.We consider the vector (T, L).

The purpose of the present section is the establishment properties and applications in risk management of the vector (T, L).

The following result establishes sufficient conditions for evaluating the distribution function of the vector (T, L).

Theorem 2.17.1 *Let N be a discrete random variable with values in the set $\mathbf{N} = \{1, 2, \ldots\}$ and probability generating function*

$$P_N(z). \tag{2.17.1}$$

We suppose that $\{X_n, n = 1, 2, \ldots\}$ is a sequence of continuous, positive, independent, and identically distributed random variables random variables. We also suppose that the random variables of the sequence are equally distributed with the random variable X having distribution function

$$F_X(x) \tag{2.17.2}$$

and we set $T = \min(X_1, X_2, \ldots, X_N)$.

Let $\{C_n, n = 1, 2, \ldots\}$ be a sequence of continuous, positive, independent, and identically distributed random variables. The random variables of the sequence are equally distributed with the random variable random variables C having distribution function

$$F_C(c) \tag{2.17.3}$$

and we set $L = \min(C_1, C_2, \ldots, C_N)$.

We consider the vector (T, L).

If N, $\{X_n, n = 1, 2, \ldots\}$ and $\{C_n, n = 1, 2, \ldots\}$ are independent then the distribution function $F_{T,L}(t, \ell)$ of the vector (T, L) is

$$F_{T,L}(t, \ell) = 1 - P_N(1 - F_X(t)) - P_N(1 - F_C(\ell)) + P_N[(1 - F_X(t))(1 - F_C(\ell))].$$

Proof The independence of N, $\{X_n, n=1,2,\ldots\}$ and $\{C_n, n=1,2,\ldots\}$ implies the independence of N and $\{X_n, n=1,2,\ldots\}$ and the independence of N and $\{C_n, n=1,2,\ldots\}$.

Hence (2.17.1) and (2.17.2) imply that the distribution function of the random variable $T=\min(X_1,X_2,\ldots,X_N)$ is

$$F_T(t) = 1 - P_N(1-F_X(t)) \tag{2.17.4}$$

and (2.17.1), (2.17.3) imply that the distribution function of the random variable $L=\min(C_1,C_2,\ldots,C_N)$ is

$$F_L(\ell) = 1 - P_N(1-F_C(\ell)). \tag{2.17.5}$$

Let $F_{T,L}(t,\ell)$ be the distribution function of the vector (T,L).We have

$$F_{T,L}(t,\ell) = P(T \le t, L \le \ell). \tag{2.17.6}$$

Since

$$P(T \le t) = P(T \le t, L \le \ell) + P(T \le t, L > \ell)$$

or equivalently

$$P(T \le t, L \le \ell) = P(T \le t) - P(T \le t, L > \ell) \tag{2.17.7}$$

and

$$P(L \ge \ell) = P(T \ge t, L \ge \ell) + P(T < t, L \ge \ell)$$

or equivalently

$$P(T < t, L \ge \ell) = P(L \ge \ell) - P(T \ge t, L \ge \ell) \tag{2.17.8}$$

then from (2.17.7) and (2.17.8) we get that

$$P(T \le t, L \le \ell) = P(T \le t) - P(L \ge \ell) + P(T \ge t, L \ge \ell). \tag{2.17.9}$$

From (2.17.4), (2.17.5), (2.17.6) and (2.17.9) it follows that

$$\begin{aligned} F_{T,L}(t,\ell) = 1 &- P_N(1-F_X(t)) - P_N(1-F_C(\ell)) \\ &+ \sum_{n=1}^{\infty} P[\min(X_1,\ldots,X_N) \ge t, \min(C_1,\ldots,C_N) \ge \ell | N=n] P(N=n). \end{aligned}$$

Hence

$$F_{T,L}(t,\ell) = 1 - P_N(1 - F_X(t)) - P_N(1 - F_C(\ell)) + \sum_{n=1}^{\infty} P[\min(X_1,\ldots,X_n) \geq t, \min(C_1,\ldots,C_n) \geq \ell | N = n] P(N = n). \quad (2.17.10)$$

From (2.17.10) it follows that

$$F_{T,L}(t,\ell) = 1 - P_N(1 - F_X(t)) - P_N(1 - F_C(\ell)) + \sum_{n=1}^{\infty} P(X_1 \geq t,\ldots,X_n \geq t, C_1 \geq \ell,\ldots,C_n \geq \ell | N = n) P(N = n). \quad (2.17.11)$$

The independence of N, $\{X_n, n = 1,2,\ldots\}$ and $\{C_n, n = 1,2,\ldots\}$ implies the independence of the random variables $N, X_1,\ldots,X_n$, $C_1,\ldots,C_n$. Hence (2.17.11) has the form

$$F_{T,L}(t,\ell) = 1 - P_N(1 - F_X(t)) - P_N(1 - F_C(\ell)) + \sum_{n=1}^{\infty} P(X_1 \geq t,\ldots,X_n, C_1 \geq \ell,\ldots,C_n \geq \ell) P(N = n). \quad (2.17.12)$$

Since the independence of the random variables N, $X_1,\ldots,X_n$, $C_1,\ldots,C_n$ implies the independence of the random variables $X_1,\ldots,X_n$, $C_1,\ldots,C_n$ then (2.17.12) has the form

$$F_{T,L}(t,\ell) = 1 - P_N(1 - F_X(t)) - P_N(1 - F_C(\ell)) + \sum_{n=1}^{\infty} P(X_1 \geq t)\ldots P(X_n \geq t) P(C_1 \geq \ell)\ldots P(C_n \geq \ell) P(N = n).$$

Since the random variables of the sequence $\{X_n, n = 1,2,\ldots\}$ are equally distributed with the random variable X having distribution function $F_X(x)$ and the random variables of the sequence $\{C_n, n = 1,2,\ldots\}$ are equally distributed with the random variable C having distribution function $F_C(c)$ then (2.17.12) has the form

$$F_{T,L}(t,\ell) = 1 - P_N(1 - F_X(t)) - P_N(1 - F_C(\ell)) + \sum_{n=1}^{\infty} (1 - F_X(t))^n (1 - F_C(\ell))^n P(N = n)$$

or equivalently

$$F_{T,L}(t,\ell) = 1 - P_N(1 - F_X(t)) - P_N(1 - F_C(\ell)) - P_N[(1 - F_X(t))(1 - F_C(\ell))].$$

An interpretation of the vector (T, L), where $T = \min(X_1, X_2,\ldots,X_N)$ and $L - \min(C_1, C_2,\ldots,C_N)$, in risk management is the following.

We consider a firm at the time point 0. The random variable N denotes the number of risk threatening the firm. The random variable X_n denotes the occurrence time of the nth risk. Hence the random variable $T = \min(X_1, X_2, \ldots, X_N)$ denotes the minimum risk occurrence time. The random variable C_n denotes the severity of the nth risk. Hence the random variable $L = \min(C_1, C_2, \ldots, C_N)$ denotes the minimum severity of risks. The vector (T, L) is a strong analytical tool for investigating the evolution of the firm under a random number N of independent competing risks. The independence of N, $\{X_n, n = 1, 2, \ldots\}$ and $\{C_n, n = 1, 2, \ldots\}$ permits the evaluation of the distribution function

$$F_{T,L}(t, \ell) = 1 - P_N(1 - F_X(t)) - P_N(1 - F_C(\ell)) + P_N[(1 - F_X(t))(1 - F_C(\ell))]$$

of the vector (T, L).

In this case the practical applicability in risk management of the above random vector is substantially extended. If the N risks threatening the firm at the time point 0 are catastrophic then the random variable $T = \min(X_1, X_2, \ldots, X_N)$, the random variable $L = \min(C_1, C_2, \ldots, C_N)$, the vector (T, L), and the corresponding distribution function

$$F_{T,L}(t, \ell) = 1 - P_N(1 - F_X(t)) - P_N(1 - F_C(\ell)) + P_N[(1 - F_X(t))(1 - F_C(\ell))]$$

constitute structural factors for investigating the evolution of the firm. □

2.18 Pair of Systems Under Competing Risks

Let $\{X_n, n = 1, 2, \ldots\}$ be a sequence of continuous, positive, independent, and identically distributed random variables. The random variables of the sequence $\{X_n, n = 1, 2, \ldots\}$ are equally distributed with the random variable X having distribution function $F_X(x)$.

We consider the discrete random variable N with values in the set $\mathbf{N} = \{1, 2, \ldots\}$ and probability generating function $P_N(z)$, and we set $T = \min(X_1, X_2, \ldots, X_N)$.

Let $\{C_s, s = 1, 2, \ldots\}$ be a sequence of continuous, positive, independent, and identically distributed random variables. The random variables of the sequence $\{C_s, s = 1, 2, \ldots\}$ are equally distributed with the random variable C having distribution function $F_C(c)$. We consider the discrete random variable S with values in the set $\mathbf{N} = \{1, 2, \ldots\}$ and probability generating function $P_S(z)$, and we set $L = \min(C_1, C_2, \ldots, C_S)$.

We consider the vector (T, L).

The purpose of the present section is the establishment of properties and applications in risk management of the above vector.

The following result establishes sufficient conditions for evaluating the distribution function of the vector (T, L).

Theorem 2.18.1 *Let $\{X_n, n = 1, 2, \ldots\}$ be a sequence of continuous, positive, independent, and identically distributed random variables. The random variables of the sequence $\{X_n, n = 1, 2, \ldots\}$ are equally distributed with the random variable X having distribution function*

$$F_X(x). \tag{2.18.1}$$

We consider the discrete random variable N with values in the set $\mathbf{N} = \{1, 2, \ldots\}$ and probability generating function

$$P_N(z) \tag{2.18.2}$$

and we set $T = \min(X_1, X_2, \ldots, X_N)$. Let $\{C_s, s = 1, 2, \ldots\}$ be a sequence of continuous, positive, independent, and identically distributed random variables. The random variables of the sequence $\{C_s, s = 1, 2, \ldots\}$ are equally distributed with the random variable C having distribution function

$$F_C(c). \tag{2.18.3}$$

We consider the discrete random variable S with values in the set $\mathbf{N} = \{1, 2, \ldots\}$, and probability generating function

$$P_S(z) \tag{2.18.4}$$

and we set $L = \min(C_1, C_2, \ldots, C_S)$.

If $\{X_n, n = 1, 2, \ldots\}$, N, $\{C_s, s = 1, 2, \ldots\}$ and S are independent then the random variables $T = \min(X_1, X_2, \ldots, X_N)$ and $L = \min(C_1, C_2, \ldots, C_S)$ are independent and

$$F_{T,L}(t, \ell) = [1 - P_N(1 - F_X(t))][1 - P_S(1 - F_C(\ell))]$$

is the distribution function of the vector (T, L).

Proof The independence of $\{X_n, n = 1, 2, \ldots\}$, N, $\{C_s, s = 1, 2, \ldots\}$ and S implies the independence of $\{X_n, n = 1, 2, \ldots\}$ and N, and the independence of $\{C_s, s = 1, 2, \ldots\}$ and S.

Hence (2.18.1) and (2.18.2) imply that the distribution function of the random variable $T = \min(X_1, X_2, \ldots, X_N)$ is

$$F_T(t) = 1 - P_N(1 - F_X(t)) \tag{2.18.5}$$

and (2.18.3), (2.18.4) imply that the distribution function of the random variable $L = \min(C_1, C_2, \ldots, C_S)$ is

$$F_L(\ell) = 1 - P_S(1 - F_C(\ell)). \tag{2.18.6}$$

Let $F_{T,L}(t, \ell)$ be the distribution function of the vector (T, L).

The establishment of the independence of the random variables $T = \min(X_1, X_2, \ldots, X_N)$ and $L = \min(C_1, C_2, \ldots, C_S)$ requires the establishment of the relationship $F_{T,L}(t,\ell) = F_T(t)F_L(\ell)$. We have

$$F_{T,L}(t,\ell) = P(T \leq t, L \leq \ell). \tag{2.18.7}$$

Since

$$P(T \leq t) = P(T \leq t, L \leq \ell) + P(T \leq t, L > \ell)$$

or equivalently

$$P(T \leq t, L \leq \ell) = P(T \leq t) - P(T \leq t, L > \ell) \tag{2.18.8}$$

and

$$P(L \geq \ell) = P(T \geq t, L \geq \ell) + P(T < t, L \geq \ell)$$

or equivalently

$$P(T < t, L \geq \ell) = P(L \geq \ell) - P(T \geq t, L \geq \ell) \tag{2.18.9}$$

then (2.18.8) and (2.18.9) imply that

$$P(T \leq t, L \leq \ell) = P(T \leq t) - P(L \geq \ell) + P(T \geq t, L > \ell) \tag{2.18.10}$$

From (2.18.5), (2.18.6), (2.18.7) and (2.18.10) it follows that

$$F_{T,L}(t,\ell) = 1 - P_N(1 - F_X(t)) - P_S(1 - F_C(\ell)) + P(T \geq t, L \geq \ell) \tag{2.18.11}$$

Moreover (2.18.11) implies that

$$\begin{aligned} F_{T,L}(t,\ell) &= 1 - P_N(1 - F_X(t)) - P_S(1 - F_C(\ell)) \\ &+ \sum_{n=1}^{\infty} \sum_{s=1}^{\infty} P\left[\min(X_1, \ldots, X_N) \geq t, \min(C_1, \ldots, C_S) \geq \ell | N = n, S = s\right] P(N = n, S = s). \end{aligned}$$

Hence

$$\begin{aligned} F_{T,L}(t,\ell) &= 1 - P_N(1 - F_X(t)) - P_S(1 - F_C(\ell)) \\ &+ \sum_{n=1}^{\infty} \sum_{s=1}^{\infty} P\left[\min(X_1, \ldots, X_n) \geq t, \min(C_1, \ldots, C_s) \geq \ell | N = n, S = s\right] P(N = n, S = s). \end{aligned} \tag{2.18.12}$$

From (2.18.12) it follows

$$F_{T,L}(t,\ell) = 1 - P_N(1 - F_X(t)) - P_S(1 - F_C(\ell))$$
$$+ \sum_{n=1}^{\infty} \sum_{s=1}^{\infty} P(X_1 \geq t, \ldots, X_n \geq t, C_1 \geq \ell, \ldots, C_s \geq \ell | N = n, S = s) P(N = n, S = s). \tag{2.18.13}$$

The independence of $\{X_n, n = 1, 2, \ldots\}$, N, $\{C_s, s = 1, 2, \ldots\}$ and S implies the independence of the random variables $X_1, \ldots, X_n$, N, $C_1, \ldots, C_s$, S and the independence of the random variables N, S.

Hence (2.18.13) has the form

$$F_{T,L}(t,\ell) = 1 - P_N(1 - F_X(t)) - P_S(1 - F_C(\ell))$$
$$+ \sum_{n=1}^{\infty} \sum_{s=1}^{\infty} P(X_1 \geq t, \ldots, X_n \geq t, C_1 \geq \ell, \ldots, C_s \geq \ell) P(N = n) P(S = s). \tag{2.18.14}$$

Since the independence of the random variables $X_1, \ldots, X_n$, N, $C_1, \ldots, C_s$, S implies the independence of the random variables $X_1, \ldots, X_n$, $C_1, \ldots, C_s$, then (2.18.14) has the form

$$F_{T,L}(t,\ell) = 1 - P_N(1 - F_X(t)) - P_S(1 - F_C(\ell))$$
$$+ \sum_{n=1}^{\infty} \sum_{s=1}^{\infty} P(X_1 \geq t) \ldots P(X_n \geq t) P(C_1 \leq \ell) \ldots P(C_s \geq \ell) P(N = n) P(S = s).$$

or equivalently the form

$$F_{T,L}(t,\ell) = 1 - P_N(1 - F_X(t)) - P_S(1 - F_C(\ell))$$
$$+ \sum_{n=1}^{\infty} P(X_1 \geq t) \ldots P(X_n \geq t) P(N = n) \sum_{s=1}^{\infty} P(C_1 \geq \ell) \ldots P(C_s \geq \ell) P(S = s). \tag{2.18.15}$$

Since the random variables of the sequence $\{X_n, n = 1, 2, \ldots\}$ are equally distributed with the random variable X having distribution function $F_X(x)$ and the random variables of the sequence $\{C_s, s = 1, 2, \ldots\}$ are equally distributed with the random variable C having distribution function $F_C(c)$ then (2.18.15) has the form

$$F_{T,L}(t,\ell) =$$
$$1 - P_N(1 - F_X(t)) - P_S(1 - F_C(\ell)) + \sum_{n=1}^{\infty} (1 - F_X(t))^n P(N = n) \sum_{s=1}^{\infty} (1 - F_C(\ell))^s P(S = s).$$

or equivalently

$$F_{T,L}(t,\ell) = 1 - P_N(1 - F_X(t)) - P_S(1 - F_C(\ell)) + P_N(1 - F_X(t)) P_S(1 - F_C(\ell)). \tag{2.18.16}$$

From (2.18.16) it follows that

$$F_{T,L}(t,\ell) = [1 - P_N(1 - F_X(t))][1 - P_S(1 - F_C(\ell))] \tag{2.18.17}$$

Moreover (2.18.5), (2.18.6) and (2.18.17) imply that $F_{T,L}(t,\ell) = F_T(t)F_L(\ell)$.

Hence the random variable $T = \min(X_1, X_2, \ldots, X_N)$ is independent of the random variable $L = \min(C_1, C_2, \ldots, C_S)$ and the distribution function $F_{T,L}(t,\ell)$ of the vector (T, L) is

$$F_{T,L}(t,\ell) = [1 - P_N(1 - F_X(t))]\,[1 - P_S(1 - F_C(\ell))]$$

An interpretation of the vector (T, L), with $T = \min(X_1, X_2, \ldots, X_N)$ and $L = \min(C_1, C_2, \ldots, C_S)$ in risk management is the following.

We consider two systems at the time point 0. The random variable N denotes the number of risks threatening the first system and the random variable S denotes the number of systems threatening the second system. The random variable X_n denotes the occurrence time of the nth risk threatening the first system. Hence the random variable $T = \min(X_1, X_2, \ldots, X_N)$ denotes the minimum risk occurrence time for the first system. The random variable C_s denotes the occurrence time of the sth risk threatening the second system. Hence the random variable $L = \min(C_1, C_2, \ldots, C_S)$ denotes the minimum risk occurrence time for the second system. The vector (T, L) is a strong analytical tool for investigating the evolution of the above mentioned systems. The independence of $\{X_n, n = 1, 2, \ldots\}$, N, $\{C_s, s = 1, 2, \ldots\}$ and S permits the evaluation of the distribution function

$$F_{T,L}(t,\ell) = [1 - P_N(1 - F_X(t))]\,[1 - P_S(1 - F_C(\ell))]$$

of the vector (T, L).

In this case the applicability of the above vector in risk management is substantially extended. □

2.19 Time of First Major Damage and Asset Liquidation

Let N be a discrete random variable with values in the set $\mathbf{N} = \{1, 2, \ldots\}$.We suppose that the random variable N follows the geometric type II distribution with probability generating function $P_N(z) = \frac{pz}{1-qz}$ and $\{C_n, n = 1, 2, \ldots\}$ is a sequence of continuous, positive, independent, and identically distributed random variables. The random variables of the sequence are equally distributed with the random variable C having characteristic function $\varphi_C(u)$. We set $Y = C_1 + C_2 + \cdots + C_N$. Let $\{\Pi_n, n = 1, 2, \ldots\}$ be a sequence of continuous, positive, independent, and identically distributed random variables. The random variables of the sequence are equally distributed with the random variable Π having characteristic function $\varphi_\Pi(\xi)$ and we set $V = \Pi_1 + \Pi_2 + \cdots + \Pi_N$. We consider the vector (Y, V).

The purpose of the present section is the establishment of properties and applications in risk management of the vector (Y,V). The following result establishes sufficient conditions for evaluating the characteristic function $\varphi_{Y,V}(u,\xi)$ of the vector (Y,V).

Theorem 2.19.1 *Let N be a discrete random variable with values in the set $\mathbf{N}=\{1,2,\ldots\}$. We suppose that the random variable N follows the geometric type II distribution with probability generating function $P_N(z)=\frac{pz}{1-qz}$ and $\{C_n, n=1,2,\ldots\}$ is a sequence of continuous, positive, independent, and identically distributed random variables. The random variables of the sequence are equally distributed with the random variable C having characteristic function $\varphi_C(u)$. We set $Y=C_1+C_2+\cdots+C_N$.*

Let $\{\Pi_n, n=1,2,\ldots\}$ be a sequence of continuous, positive, independent, and identically distributed random variables. The random variables of the sequence are equally distributed with the random variable Π having characteristic function $\varphi_\Pi(\xi)$ and we set $V=\Pi_1+\Pi_2+\cdots+\Pi_N$. If N, $\{C_n, n=1,2,\ldots\}$ and $\{\Pi_n, n=1,2,\ldots\}$ are independent then the characteristic function of the vector (Y,V) is

$$\varphi_{Y,V}(u,\xi)=\frac{p\varphi_C(u)\varphi_\Pi(\xi)}{1-q\varphi_C(u)\varphi_\Pi(\xi)}$$

Proof The proof of Theorem 2.19.1 follows from the proof of Theorem 2.12.1. An interpretation of vector (Y,V), where $Y=C_1+C_2+\cdots+C_N$ and $V=\Pi_1+\Pi_2+\cdots+\Pi_N$, in risk management is the following.

A firm faces a risk. We suppose that $\{X_n, n=1,2,\ldots\}$ is a sequence of continuous, positive, independent, and identically distributed random variables. The random variables of the sequence are equally distributed with the random variable X having distribution function

$$F_X(x). \tag{2.19.1}$$

The random variable X_n denotes the size of the damage due to the nth occurrence of the risk threatening the firm.

Let θ be a positive real number. If $X_n>\theta$ then the damage due to the nth risk occurrence is considered as a major one. If p is the probability of the event "the damage due the nth risk occurrence is major", then $p=P(X_n>\theta)$ or equivalently $p=1-P(X_n\leq\theta)$.

Hence (2.19.1) implies that $p=1-F_X(\theta)$. If the random variable N denotes the number of risk occurrences required for the first appearance of a major damage then the random variable N follows the geometric type II distribution with probability generating function $P_N(z)=\frac{pz}{1-qz}$.

We consider the sequence of random variables $\{C_n, n=1,2,\ldots\}$ and we suppose that the random variable C_n denotes the time between the nth and the $(n-1)$th risk occurrence. Hence the random variable $Y = C_1 + C_2 + \cdots + C_N$ denotes the time of the first appearance of a major damage.

We consider the sequence of random variables $\{\Pi_n, n = 1, 2, \ldots\}$ and we suppose that the random variable Π_n denotes the income of the firm from asset liquidation at the time point of the nth risk occurrence. Hence the random variable $V = \Pi_1 + \Pi_2 + \cdots + \Pi_N$ denotes the total income of the firm from asset liquidation in the time interval $[0, Y]$.

In this case the vector (Y, V) is a strong analytic tool for investigating the evolution of the firm under conditions of the first appearance of a major damage $Y = C_1 + C_2 + \cdots + C_N$ and the total income $V = \Pi_1 + \Pi_2 + \cdots + \Pi_N$ obtained by asset liquidation until the time point $Y = C_1 + C_2 + \cdots + C_N$.

The independence for N, $\{C_n, n = 1, 2, \ldots\}$ and $\{\Pi_n, n = 1, 2, \ldots\}$ permits the evaluation of the characteristic function

$$\varphi_{Y,V}(u, \xi) = \frac{p\varphi_C(u)\varphi_\Pi(\xi)}{1 - q\varphi_C(u)\varphi_\Pi(\xi)}$$

of the vector (Y, V).

In this case the practical applicability of the above vector in risk management is substantially extended. □

2.20 Time of First Major Damage and Loan Portfolio

Let $\{C_n, n = 1, 2, \ldots\}$ be a sequence of continuous, positive, independent, and identically distributed random variables. The random variables of the sequence are equally distributed with the random variable C having characteristic function $\varphi_C(u)$.

We suppose that N is a discrete random variable with values in the set $\mathbf{N} = \{1, 2, \ldots\}$ and the random variable N follows the geometric type II distribution with probability generating function $P_N(z) = \frac{pz}{1-qz}$ and we set $Y = C_1 + C_2 + \cdots + C_N$.

Let $\{\Pi_s, s = 1, 2, \ldots\}$ be a sequence of continuous, positive, independent, and identically distributed random variables. The random variables of the sequence are equally distributed with the random variable Π having characteristic function $\varphi_\Pi(\xi)$.

We suppose that S is a discrete random variable with values in the set $\mathbf{N}_0 = \{0, 1, 2, \ldots\}$ and probability generating function $P_S(z)$ and we set $V = \Pi_1 + \Pi_2 + \cdots + \Pi_S$.

We consider the vector (Y, V).

The purpose of the present section is the establishment of properties and applications in risk management of the vector (Y, V).

The following result establishes sufficient conditions for evaluating the characteristic function $\varphi_{Y,V}(u,\xi)$ of the vector (Y,V).

Theorem 2.20.1 *Let $\{C_n, n = 1,2,\ldots\}$ be a sequence of continuous, positive, independent, and identically distributed random variables. The random variables of the sequence are equally distributed with the random variable C.*

We suppose that N is a discrete random variable with values in the set $\mathbf{N} = \{1,2,\ldots\}$ and the random variable N follows the geometric type II distribution with probability generating function $P_N(z) = \frac{pz}{1-qz}$ and we set $Y = C_1 + C_2 + \cdots + C_N$.

Let $\{\Pi_s, s = 1,2,\ldots\}$ be a sequence of continuous, positive, independent, and identically distributed random variables. The random variables of the sequence are equally distributed with the random variable Π having characteristic function $\varphi_\Pi(\xi)$. We suppose that S is a discrete random variable with values in the set $\mathbf{N}_0 = \{0,1,2,\ldots\}$ and probability generating function $P_S(z)$ and we set $V = \Pi_1 + \Pi_2 + \cdots + \Pi_S$. If $\{C_n, n = 1,2,\ldots\}$, N, $\{\Pi_s, s = 1,2,\ldots\}$, and S are independent then the random variables $Y = C_1 + C_2 + \cdots + C_N$ and $V = \Pi_1 + \Pi_2 + \cdots + \Pi_S$ are independent and

$$\varphi_{Y,V}(u,\xi) = \frac{p\varphi_C(u)}{1 - q\varphi_C(u)} P_S(\varphi_\Pi(\xi))$$

is the characteristic function of the vector (Y,V).

Proof The proof of Theorem 2.20.1 follows from the proof of Theorem 2.13.1. An interpretation of the vector (Y,V), where $Y = C_1 + C_2 + \cdots + C_N$ and $V = \Pi_1 + \Pi_2 + \cdots + \Pi_S$, in risk management is the following.

A bank faces a risk. We suppose that $\{X_n, n = 1,2,\ldots\}$ is a sequence of continuous, positive, independent, and identically distributed random variables. The random variables of the sequence are equally distributed with the random variable X having distribution function

$$F_X(x) \tag{2.20.1}$$

The random variable X_n denotes the damage due to the nth occurrence of the risk threatening the bank.

Let θ be a positive real number. If $X_n > \theta$ then the damage due to the nth risk occurrence is considered as major one. If p is the probability of the event "the damage due to the nth risk occurrence is major", then $p = P(X_n > \theta)$ or equivalently $p = 1 - P(X_n \le \theta)$.

Hence (2.20.1) implies that $p = 1 - F_X(\theta)$.

If the random variable N denotes the number of risk occurrences required for the first appearance of a major damage then the random variable N follows the geometric type II distribution with probability generating function $P_N(z) = \frac{pz}{1-qz}$.

We consider the sequence of random variables $\{C_n, n = 1, 2, \ldots\}$ and we suppose that the random variable C_n denotes the time between the nth and the $(n-1)$th risk occurrence. Hence the random variable $Y = C_1 + C_2 + \cdots + C_N$ denotes the time of the first appearance of a major damage. We consider the discrete random variable S denoting the number of loans in the portfolio of loans of the bank at the time point $Y = C_1 + C_2 + \cdots + C_N$.

We consider the sequence of random variables $\{\Pi_s, s = 1, 2, \ldots\}$ and we suppose that the random variable Π_s denotes the size of the sth loan of the portfolio of loans of the bank at the time point $Y = C_1 + C_2 + \cdots + C_N$. Hence the random variable $V = \Pi_1 + \Pi_2 + \cdots + \Pi_S$ denotes the total size of the portfolio of loans of the bank at the time point $Y = C_1 + C_2 + \cdots + C_N$. In this case the vector (Y, V) is a strong analytical tool for investigating the evolution of the bank under the time of the first appearance of a major damage $Y = C_1 + C_2 + \cdots + C_N$ and the total size of the portfolio of loans of the bank at the time point $Y = C_1 + C_2 + \cdots + C_N$.

The independence for N, $\{C_n, n = 1, 2, \ldots\}$, S and $\{\Pi_s, s = 1, 2, \ldots\}$ permits the evaluation of the characteristic function

$$\varphi_{Y,V}(u, \xi) = \frac{p\varphi_C(u)}{1 - q\varphi_C(u)} P_S(\varphi_\Pi(\xi))$$

of the vector (Y, V).

In this case the practical applicability in risk management of the above vector is substantially extended. □

Chapter 3
Stochastic Models of Risk Management Operations

Abstract This chapter is devoted to the implementation of two purposes. The formulation and investigation of stochastic models for the fundamental risk treatment operations are the two purposes. Risk control operations and risk financing operations constitute the two categories of fundamental risk treatment operations. This chapter consists of two parts. The first part concentrates on stochastic modeling of risk reduction operations, risk duration reduction operations, and risk frequency reduction operations which are the main risk control operations. The second part concentrates on stochastic modeling of risk financing operations. The stochastic modeling of the cost for treatment of ongoing risk occurrences, and the reserve of risk financing constitute the main theoretical and practical contribution of the second part of this chapter. In theory and practice, risk treatment is implemented by combining stochastic models of risk control operations with stochastic models of risk financing operations.

3.1 Introduction

The purpose of the present chapter is the formulation and investigation of stochastic models for the fundamental risk treatment operations. Risk management literature recognizes two categories of fundamental risk treatment operations. The first category includes the risk control operations and the second category includes the risk financing operations. In practice, risk treatment is frequently implemented by combining risk control and risk financing operations. The first part of the present chapter concentrates on stochastic modeling of risk severity reduction operations, risk duration reduction operations, and risk frequency reduction operations. Stochastic multiplicative models incorporating two continuous, positive, and independent random variables are used for describing and analyzing risk severity reduction operations and risk duration reduction operations. The concepts and stochastic models of operations of deleting risk occurrences with constant probability, operations of deleting risk occurrences with random probability, and

C. Artikis and P. Artikis, *Probability Distributions in Risk Management Operations*, Intelligent Systems Reference Library 83, DOI 10.1007/978-3-319-14256-2_3

operations of uniform risk frequency reduction constitute the main contribution of the first part of the present chapter. The first part of this chapter includes the formulation and investigation of a stochastic multiplicative model for describing and analyzing the cost of operations of deleting risk occurrences with constant probability. The second part concentrates on the concepts and models of risk financing operations. The stochastic modeling of the cost for treatment of ongoing risk occurrences, the impacts of risks and the reserve of risk financing constitute the contribution of the second part of the present chapter.

The complexity of stochastic models of risk control operations and stochastic models of risk financing operations is significant because these models use as structural elements the stochastic models of the fundamental concepts of risk management. The complexity of such models does not create particular difficulties in risk analysts, risk managers, and risk experts.

The investigation and application of stochastic models for risk control operations and risk financing operations are strongly supported by the powerful results of the theory of characteristic functions. These results permit the establishment of important properties for the probability distributions of such stochastic models. Moreover, these results make characteristic functions very useful analytical tools of risk management.

3.2 Models of Risk Severity Reduction Operations

We consider a risk with severity represented by the continuous and positive random variable X having distribution function $F_X(x)$, probability density function $f_X(x)$, and characteristic function $\varphi_X(u)$.

An operation of reducing the severity X of a risk is a process which is applied to the conditions and the cause of that risk. The consequence of that process is the representation of the severity of the risk with a continuous and positive random variable Y such that $P(Y<X)=1$.

Since the random variable Y denotes the severity of risk after applying a risk severity reduction operation then this random variable is a function of the random variable X which denotes the severity of risk before applying the risk severity reduction operation. The formulation of that function between the random variable Y and the random variable X is called stochastic modeling of the risk severity reduction operation. The result of the stochastic modeling of the risk frequency reduction operation is the stochastic model denoting the severity of risk Y after applying the risk severity reduction operation. This model is called stochastic model of a risk severity reduction operation. A large class of such stochastic models is the class of stochastic multiplicative models of the form $Y = XU$ where U is a continuous random variable with values in the interval $(0, 1)$, distribution function $F_U(v)$, probability density function $f_U(v)$ and characteristic function $\varphi_U(u)$.

Moreover, the random variable U is independent of the random variable X. The evaluation of the distribution function $F_Y(y)$, the evaluation of the probability density function $f_Y(y)$ and the evaluation of the characteristic function $\varphi_Y(u)$ of the stochastic model $Y = XU$ are necessary for the investigation and applications of this model. Following Sect. 2.10 we get the formulas

$$F_Y(y) = \int_0^1 F_X\left(\frac{y}{\upsilon}\right) f_U(\upsilon)\, d\upsilon, \tag{3.2.1}$$

$$f_Y(y) = \int_0^1 \frac{1}{\upsilon} f_X\left(\frac{y}{\upsilon}\right) f_U(\upsilon)\, d\upsilon, \tag{3.2.2}$$

$$\varphi_Y(u) = \int_0^1 \varphi_X(u\upsilon) f_U(\upsilon)\, d\upsilon, \tag{3.2.3}$$

and the formulas

$$F_Y(y) = \int_0^\infty F_U\left(\frac{y}{x}\right) f_X(x)\, dx,$$

$$f_Y(y) = \int_0^\infty \frac{1}{x} f_U\left(\frac{y}{x}\right) f_X(x)\, dx,$$

$$\varphi_Y(u) = \int_0^\infty \varphi_U(ux) f_X(x)\, dx.$$

Particular theoretical and practical interest can be recognized in the two following special cases of the stochastic model $Y = XU$.

The first special case of the above model arises if the random variable U follows the beta distribution with probability density function

$$f_U(\upsilon) = a\upsilon^{a-1}, \quad 0 < \upsilon < 1, \;\; a > 0. \tag{3.2.4}$$

From (3.2.1), (3.2.2), (3.2.3) and (3.2.4) we get the formula

$$F_Y(y) = a \int_0^1 F_X\left(\frac{y}{\upsilon}\right) \upsilon^{a-1}\, d\upsilon$$

for the distribution function, the formula

$$f_Y(y) = a \int_0^1 \frac{1}{v} f_X\left(\frac{y}{v}\right) v^{a-1}\, dv$$

for the probability density function, and the formula

$$\varphi_Y(u) = a \int_0^1 \varphi_X(uv) v^{a-1}\, dv$$

for the characteristic function of the stochastic multiplicative model $Y = XU$.

The above formulas correspond in the class of a-unimodal probability distributions.

The following result provides an interesting interpretation of the class of a-unimodal probability distributions in the area of risk severity reduction operations.

Theorem 3.2.1 *Let X be a continuous and positive random variable with characteristic function $\varphi_X(u)$, Y a continuous and positive random variable with differentiable characteristic function $\varphi_Y(u)$ and U a random variable following the beta distribution with probability density function*

$$f_U(v) = av^{a-1}, \quad 0 < v < 1, \quad a > 0.$$

We suppose that the random variables X, Y, U are independent. The characteristic function $\varphi_Y(u)$ of the random variable Y has the form

$$\varphi_Y(u) = \exp\left(a \int_0^u \frac{\varphi_X(y) - 1}{y} dy \right)$$

if, and only if, $Y \stackrel{d}{=} (X + Y)U$ where $\stackrel{d}{=}$ means equality in distribution.

Proof The characteristic function of the random variable $L = (X + Y)U$ is

$$\varphi_L(u) = E\left(e^{iu(X+Y)U}\right). \tag{3.2.5}$$

From (3.2.5) we get that

$$\varphi_L(u) = E\left(E\left(e^{iu(X+Y)U} \,|\, U\right)\right)$$

or equivalently

$$\varphi_L(u) = a\int_0^1 E\left(e^{iu(X+Y)U}|U = \upsilon\right)\upsilon^{a-1}d\upsilon. \tag{3.2.6}$$

Moreover (3.2.6) implies that

$$\varphi_L(u) = a\int_0^1 E\left(e^{iu\upsilon(X+Y)}|U = \upsilon\right)\upsilon^{a-1}d\upsilon$$

or equivalently

$$\varphi_L(u) = a\int_0^1 E(e^{iu\upsilon X}\cdot e^{iu\upsilon Y}|U = \upsilon)\upsilon^{a-1}d\upsilon. \tag{3.2.7}$$

Since the random variables X, Y, U are independent then the random variables $e^{iu\upsilon X}$, $e^{iu\upsilon Y}$, U are also independent. Hence (3.2.7) has the form

$$\varphi_L(u) = a\int_0^1 E(e^{iu\upsilon X}\cdot e^{iu\upsilon Y})\upsilon^{a-1}d\upsilon. \tag{3.2.8}$$

The independence of the random variables X, Y, U imply the independence of the random variables X, Y and this conclusion implies the independence of the random variables $e^{iu\upsilon X}$, $e^{iu\upsilon Y}$.

Hence (3.2.8) has the form

$$\varphi_L(u) = a\int_0^1 E(e^{iu\upsilon X})E(e^{iu\upsilon Y})\upsilon^{a-1}d\upsilon. \tag{3.2.9}$$

Since $\varphi_X(u) = E(e^{iuX})$ and $\varphi_Y(u) = E(e^{iuY})$ then (3.2.9) has the form

$$\varphi_L(u) = a\int_0^1 \varphi_X(u\upsilon)\varphi_Y(u\upsilon)\upsilon^{a-1}d\upsilon. \tag{3.2.10}$$

If we introduce the characteristic function $\varphi_Y(u)$ and the characteristic function $\varphi_L(u)$, in (3.2.10), to the relationship

$$Y \stackrel{d}{=} (X+Y)U$$

then we get the integral equation

$$\varphi_Y(u) = a\int_0^1 \varphi_x(uv)\varphi_Y(uv)v^{a-1}dv. \tag{3.2.11}$$

From (3.2.11) it follows that

$$\varphi_Y(u) = \frac{a}{u^a}\int_0^u \varphi_x(y)\varphi_Y(y)y^{a-1}dy \tag{3.2.12}$$

and (3.2.12) it follows that

$$u^a\varphi_Y(u) = a\int_0^u \varphi_x(y)\varphi_Y(y)y^{a-1}dy. \tag{3.2.13}$$

If we differentiate (3.2.13) then we get the differential equation

$$au^{a-1}\varphi_Y(u) + u^a\frac{d\varphi_Y(u)}{du} = a\varphi_X(u)\varphi_Y(u)u^{a-1} \tag{3.2.14}$$

which satisfies the conditions $\varphi_Y(0) = 1$ and $\varphi_X(0) = 1$.

If $u \neq 0$ then the differential equation (3.2.14) has the form

$$\frac{d\varphi_Y(u)}{du} = a\frac{\varphi_X(u) - 1}{u}\varphi_Y(u). \tag{3.2.15}$$

From (3.2.15) we get that

$$\int_0^u \frac{d\varphi_Y(y)}{\varphi_Y(y)} = a\int_0^u \frac{\varphi_X(y) - 1}{y}dy,$$

for every y such that $\varphi_Y(y) \neq 0$.

Hence

$$\log\varphi_Y(u) = a\int_0^u \frac{\varphi_X(y) - 1}{y}dy$$

or equivalently

$$\varphi_Y(u) = \exp\left(a\int_0^u \frac{\varphi_X(y) - 1}{y}dy\right). \tag{3.2.16}$$

Conversely, we suppose that (3.2.16) is valid. We have

$$\exp\left(a\int_0^u \frac{\varphi_X(y)-1}{y}dy\right) = \frac{a}{u^a}\int_0^u \varphi_X(y)\exp\left(a\int_0^u \frac{\varphi_X(w)-1}{w}dy\right)y^{a-1}dy$$

or equivalently

$$\exp\left(a\int_0^u \frac{\varphi_X(y)-1}{y}dy\right) = a\int_0^1 \varphi_X(uv)\exp\left(a\int_0^{uv} \frac{\varphi_X(y)-1}{y}dy\right)v^{a-1}dv. \tag{3.2.17}$$

The independence of the random variables X, Y, U and (3.2.17) imply that $Y \stackrel{d}{=} (X+Y)U$.

If the random variable X has a finite mean then the characteristic function

$$\varphi_Y(u) = \exp\left(a\int_0^u \frac{\varphi_X(y)-1}{y}dy\right)$$

belongs to a self-decomposable probability distribution. Since the self-decomposable probability distributions are unimodal then the above result is of particular practical interest.

An interpretation of the above theoretical result in the area of risk frequency reduction operations is the following. We consider two independent risks of the same kind. We suppose that the continuous and positive random variable X denotes the severity of one risk and the random variable Y denotes the severity of the other risk. Moreover, the same risk severity reduction operation is applied to these risks. The same kind of risks and the application of the same risk severity reduction operation permit the consideration of the two risks as a whole. We suppose that the random variable U with probability density function

$$f_U(v) = av^{a-1}, \quad 0<v<1, \ a>0,$$

denotes the impact of the application of the risk severity reduction operation to the total severity $X+Y$ of the two risks. Hence the random variable $(X+Y)U$ denotes the total severity of the two risks after the application of the risk severity reduction operation. We suppose that the random variables X, Y, U are independent. The equality in distribution of the random variables Y and $(X+Y)U$ implies that the characteristic function $\varphi_Y(u)$ of the random variable Y has the form

$$\varphi_Y(u) = \exp\left(a \int_0^u \frac{\varphi_X(y) - 1}{y} dy\right)$$

and conversely the consideration of the characteristic function

$$\varphi_Y(u) = \exp\left(a \int_0^u \frac{\varphi_X(y) - 1}{y} dy\right)$$

as characteristic function of the random variable Y implies the equality in distribution of the random variables Y and $(X+Y)U$.

The practical significance of the above result is supported by the assumption that the existence of a finite mean for the random variable X implies the consideration of the characteristic function

$$\varphi_Y(u) = \exp\left(a \int_0^u \frac{\varphi_X(y) - 1}{y} dy\right)$$

as characteristic function of a self-decomposable probability distribution. In this case the above result establishes a relationship between the class of self-decomposable distributions and the class of a-unimodal probability distributions. This means that the class of self-decomposable probability distributions can be used for the investigation of stochastic multiplicative models of risk severity reduction operations.

If $a<1$ then the random variable U has probability density function $f_U(\upsilon) = a\upsilon^{a-1}, 0<\upsilon<1$ which is unimodal at the point 0. Under the assumption $a<1$ the result of Medgyessy implies that the probability density function

$$f_Y(y) = a \int_0^1 \frac{1}{\upsilon} f_X\left(\frac{y}{\upsilon}\right) \upsilon^{a-1} d\upsilon$$

of the stochastic model $Y = XU$ has a unique mode at the point 0.

The second special case of the model $Y = XU$ arises if the random variable U follows the beta distribution with probability density function

$$f_U(\upsilon) = \nu(1-\upsilon)^{\nu-1}, \quad 0<\upsilon<1, \quad \nu > 0. \tag{3.2.18}$$

From (3.2.1), (3.2.2), (3.2.3) and (3.2.18) we get the formula

$$F_Y(y) = \nu \int_0^1 F_X\left(\frac{y}{\upsilon}\right) (1-\upsilon)^{\nu-1} d\upsilon$$

for the distribution function, the formula

$$f_Y(y) = \nu \int_0^1 \frac{1}{\upsilon} f_X\left(\frac{y}{\upsilon}\right)(1-\upsilon)^{\nu-1} d\upsilon$$

for the probability density function and the formula

$$\varphi_Y(u) = \nu \int_0^1 \varphi_X(u\upsilon)(1-\upsilon)^{\nu-1} d\upsilon$$

for the characteristic function of the stochastic model $Y = XU$.

The above formulas correspond to the class of ν-unimodal probability distributions.

If $\nu > 1$ then the random variable U has probability density function

$$f_U(\upsilon) = \nu(1-\upsilon)^{\nu-1}, \quad 0 < \upsilon < 1$$

which is unimodal at the point 0. If $\nu > 1$ then the result of Medgyessy implies that the probability density function

$$f_Y(y) = \nu \int_0^1 \frac{1}{\upsilon} f_X\left(\frac{y}{\upsilon}\right)(1-\upsilon)^{\nu-1} d\upsilon$$

of the stochastic model $Y = XU$ has a unique mode at the point 0.

If $a = 1$ or $\nu = 1$ then it easily follows the formula

$$F_Y(y) = \int_0^1 F_X\left(\frac{y}{\upsilon}\right) d\upsilon$$

for the distribution function, the formula

$$f_Y(y) = \int_0^1 \frac{1}{\upsilon} f_X\left(\frac{y}{\upsilon}\right) d\upsilon$$

for the probability density function and the formula

$$\varphi_Y(u) = \int_0^1 \varphi_X(uv)dv$$

for the characteristic function of the stochastic model $Y = XU$.

If $a = 1$ or $v = 1$ the result of Khintchine implies that the probability density function

$$f_Y(y) = \int_0^1 \frac{1}{v} f_X\left(\frac{y}{v}\right) dv$$

of the stochastic model $Y = XU$ has a unique mode at the point 0.

The existence of a unique mode at the point 0 for the probability density function $f_Y(y)$ of the random variable Y, which denotes the severity of a risk after the application of a risk severity reduction operation, has theoretical and practical importance for risk control operations. The theoretical importance of the unimodality at the point 0 of the probability density function $f_Y(y)$ for the risk control operations is the significant probability of the event "the risk severity Y, after the application of a risk severity reduction operation, to be in an area right to the point 0". The practical significance of unimodality at the point 0 of the probability density function $f_Y(y)$ for the risk control operations is the effectiveness of the risk severity reduction operation which transforms the random variable X into the random variable Y.

The contribution of the present section consists of introducing a class of stochastic multiplicative models of the form $Y = XU$ for the description and analysis of risk severity reduction operations. The main advantage of this class is the inclusion of stochastic models, related with the study of the property of unimodality at the point 0 of probability density functions, in this class. The presence of the continuous random variable U, with values in the interval $(0, 1)$, in the stochastic multiplicative models of the form $Y = XU$ is a structural factor in introducing a unique mode at the point 0 for the probability density function $f_Y(y)$ of the random variable Y.

The interpretation of the random variable U as a coefficient of the effectiveness of a risk severity reduction operation and the consideration of the random variable U as a structural factor in introducing a unique mode at the point 0 for the probability density function $f_Y(y)$ of the stochastic model $Y = XU$ result in the conclusion that stochastic models based on the product of two independent and positive random variables, one of which takes values in the interval $(0, 1)$, constitute strong analytical tools for investigating effective risk severity reduction operations. □

3.3 Models of Risk Duration Reduction Operations

We consider a risk with duration denoted by the continuous and positive random variable S having distribution function $F_S(s)$, probability density function $f_S(s)$, and characteristic function $\varphi_S(u)$.

An operation reducing the duration S of a risk is a process which is applied to the conditions and the cause of a risk. The consequence of that process is the representation of the risk duration with a continuous and positive random variable Y such that $P(Y < S) = 1$.

Since the random variable Y denotes the duration of risk after applying a risk duration reduction operation then this random variable is a function of the random variable S which denotes the duration of risk before applying the risk duration reduction operation. The formulation of the function between the random variable Y and the random variable S is called stochastic modeling of the risk duration reduction operation. The result of stochastic modeling of the risk duration reduction operation is the stochastic model denoting the risk duration Y after applying the risk duration reduction operation. This model is called stochastic model of a risk duration reduction operation. A large class of such stochastic models is the class of stochastic multiplicative models of the form $Y = SU$ where U is a continuous random variable with values in the interval $(0, 1)$, distribution function $F_U(\upsilon)$, probability density function $f_U(\upsilon)$, and characteristic function $\varphi_U(u)$.

Moreover, the random variable U is independent of the random variable S.

The evaluation of the distribution function $F_Y(y)$, probability density function $f_Y(y)$, and characteristic function $\varphi_Y(u)$ of the stochastic model $Y = SU$ are required for the investigation and the applications of that model. Following Sect. 2.10 we get the formulas

$$F_Y(y) = \int_0^1 F_S\left(\frac{y}{\upsilon}\right) f_U(\upsilon) d\upsilon, \tag{3.3.1}$$

$$f_Y(y) = \int_0^1 \frac{1}{\upsilon} f_S\left(\frac{y}{\upsilon}\right) f_U(\upsilon) d\upsilon, \tag{3.3.2}$$

$$\varphi_Y(u) = \int_0^1 \varphi_S(u\upsilon) f_U(\upsilon) d\upsilon \tag{3.3.3}$$

and the formulas

$$F_Y(y) = \int_0^\infty F_U\left(\frac{y}{s}\right) f_S(s)ds,$$

$$f_Y(y) = \int_0^\infty \frac{1}{s} f_U\left(\frac{y}{s}\right) f_S(s)ds,$$

$$\varphi_Y(u) = \int_0^\infty \varphi_U(us) f_S(s)ds.$$

Particular theoretical and practical interest can be recognized in the two following special cases of the stochastic model $Y = SU$.

The first special case is the model which arises if the random variable U follows the beta distribution with probability density function

$$f_U(\upsilon) = a\upsilon^{a-1}, \quad 0 < \upsilon < 1, \ a > 0. \tag{3.3.4}$$

From (3.3.1), (3.3.2), (3.3.3) and (3.3.4) we get the formula

$$F_Y(y) = a \int_0^1 F_S\left(\frac{y}{\upsilon}\right) \upsilon^{a-1} d\upsilon$$

for the distribution function, the formula

$$f_Y(y) = a \int_0^1 \frac{1}{\upsilon} f_S\left(\frac{y}{\upsilon}\right) \upsilon^{a-1} d\upsilon$$

for the probability density function and the formula

$$\varphi_Y(u) = a \int_0^1 \varphi_S(u\upsilon) \upsilon^{a-1} d\upsilon \tag{3.3.5}$$

for the characteristic function of the stochastic multiplicative model $Y = SU$. The above formulas correspond to the class of a-unimodal probability distributions.

If $a < 1$ then the random variable U has probability density function $f_U(\upsilon) = a\upsilon^{a-1}, 0 < \upsilon < 1$ which is unimodal at the point 0. If $a < 1$ then the result of Medgyessy implies that the probability density function

$$f_Y(y) = a\int_0^1 \frac{1}{v} f_S\left(\frac{y}{v}\right) v^{a-1}dv$$

of the stochastic model $Y = SU$ has a unique mode at the point 0. If $a = 1$ then it easily follows the formula

$$F_Y(y) = \int_0^1 F_S\left(\frac{y}{v}\right)dv$$

for the distribution function, the formula

$$f_Y(y) = \int_0^1 \frac{1}{v} f_S\left(\frac{y}{v}\right)dv$$

for the probability distribution function and the formula

$$\varphi_Y(u) = \int_0^1 \varphi_S(uv)dv$$

for the characteristic function of the stochastic model $Y = SU$.

If $a = 1$ the result of Khintchine implies that the probability density function

$$f_Y(y) = \int_0^1 \frac{1}{v} f_S\left(\frac{y}{v}\right)dv$$

of the stochastic model $Y = SU$ has a unique mode at the point 0.

The consideration of the special case of the characteristic function (3.3.5), with the random variable S following the gamma distribution, is of particular practical interest since the gamma distribution is a common distribution of time. We suppose that the random variable S, which represents the duration of risk before applying the risk duration reduction operation, follows the gamma distribution with characteristic function

$$\varphi_S(u) = \left(\frac{\mu}{\mu - iu}\right)^{a+1}. \tag{3.3.6}$$

From (3.3.5) and (3.3.6) it follows that the characteristic function of the random variable Y, representing the duration of risk after applying the risk duration reduction operation, has the form

$$\varphi_Y(u) = a\int_0^1 \left(\frac{\mu}{\mu - iuv}\right)^{a+1} v^{a-1} dv$$

or equivalently the form

$$\varphi_Y(u) = \left(\frac{\mu}{\mu - iu}\right)^a.$$

Hence the random variable Y follows the gamma distribution with parameters μ, α.

The consideration of the special case of the characteristic function (3.3.5) with the random variable S following the exponential distribution and $a = 1$ is of particular practical interest since the exponential distribution is the most common distribution of time and the standard uniform has many applications in stochastic modeling of operations of multiplicative reduction of random factors.

We suppose that the random variable S follows the exponential distribution with characteristic function

$$\varphi_S(u) = \frac{\mu}{\mu - iu} \tag{3.3.7}$$

and $a = 1$.

From (3.3.5) and (3.3.7) it follows that the characteristic function of the random variable Y has the form

$$\varphi_Y(u) = \int_0^1 \frac{\mu}{\mu - iuv} dv$$

or equivalently the form

$$\varphi_Y(u) = \frac{\mu}{iu}\log\left(\frac{\mu}{\mu - iu}\right). \tag{3.3.8}$$

The characteristic function (3.3.8) belongs to the class of infinitely divisible characteristic functions and the corresponding probability density function

$$f_Y(y) = \int_0^1 \frac{\mu}{v} e^{-\mu\frac{y}{v}} dv$$

has a unique mode at the point 0. Since it is not possible the analytic evaluation of the probability density function

$$f_Y(y) = \int_0^1 \frac{\mu}{v} e^{-\mu \frac{y}{v}} dv$$

and the analytic evaluation of the distribution function

$$F_Y(y) = \int_0^1 \left(1 - e^{-\mu \frac{y}{v}}\right) dv$$

then the investigation of risk duration $Y = SU$ is based on the characteristic function

$$\varphi_Y(u) = \frac{\mu}{iu} \log\left(\frac{\mu}{\mu - iu}\right).$$

The use of the characteristic function

$$\varphi_Y(u) = \frac{\mu}{iu} \log\left(\frac{\mu}{\mu - iu}\right)$$

for the investigation of risk duration $Y = SU$ is implemented by two methods. The first method is based on the inversion theorem for characteristic functions and the algorithm called Fast Fourier Transform. This method provides an analytic approximation of the distribution function

$$F_Y(y) = \int_0^1 \left(1 - e^{-\mu \frac{y}{v}}\right) dv$$

and an analytic approximation of the probability density function

$$f_Y(y) = \int_0^1 \frac{\mu}{v} e^{-\mu \frac{y}{v}} dv$$

of the model $Y = SU$.

The second method is the establishment of properties for the characteristic function

$$\varphi_Y(u) = \frac{\mu}{iu} \log\left(\frac{\mu}{\mu - iu}\right)$$

which imply properties for the distribution function

$$F_Y(y) = \int_0^1 \left(1 - e^{-\mu\frac{y}{v}}\right) dv$$

and properties for the probability density function

$$f_Y(y) = \int_0^1 \frac{\mu}{v} e^{-\mu\frac{y}{v}} dv.$$

This method provides a partial description of the probabilistic behaviour of the model $Y = SU$.

The integral representation

$$\varphi_Y(u) = \int_0^1 \frac{\mu}{\mu - iuv} dv$$

of the characteristic function

$$\varphi_Y(u) = \frac{\mu}{iu} \log\left(\frac{\mu}{\mu - iu}\right)$$

of the model $Y = SU$ implies the property of unimodality at the point 0 and the property of infinite divisibility of the probability density function

$$f_Y(y) = \int_0^1 \frac{\mu}{v} e^{-\mu\frac{y}{v}} dv.$$

These properties provide significant information for the behaviour of the probability density function

$$f_Y(y) = \int_0^1 \frac{\mu}{v} e^{-\mu\frac{y}{v}} dv$$

at the tails of its domain. The establishment of the characteristic function

$$\varphi_Y(u) = \frac{\mu}{iu} \log\left(\frac{\mu}{\mu - iu}\right)$$

as a member of a class of transformed infinitely divisible characteristic functions is of particular theoretical and practical interest for the risk duration reduction operations. This establishment is based on the following result of Abate and Whitt.

Let V be a continuous and positive random variable with characteristic function $\varphi_V(u)$ and finite mean m, then

$$\varphi_B(u) = \frac{1}{imu} \log \varphi_V(u) \tag{3.3.9}$$

is the characteristic function of a continuous and positive random variable B with probability density function $f_B(\beta)$ which is unimodal at the point 0 if, and only if, the random variable V is infinitely divisible. The formula (3.3.9) can be considered as transformation which converts the characteristic function $\varphi_V(u)$ of a continuous, positive, and infinitely divisible random variable with finite mean m into the characteristic function $\varphi_B(u)$ of a random variable B with probability density function $f_B(\beta)$ having unique mode at the point 0.

Theorem 3.3.1 *Let S be a continuous, positive, and infinitely divisible random variable with characteristic function $\varphi_S(u)$ and finite mean θ.*

Let Y be a continuous and positive random variable with characteristic function

$$\varphi_Y(u) = \frac{1}{i\theta u} \log \varphi_S(u).$$

If U is a random variable following the uniform distribution with probability density function $f_U(v) = 1, 0 < v < 1$ and the random variable U is independent of the random variable S then

$$\varphi_S(u) = \frac{\mu}{\mu - iu}$$

with $\mu = \frac{1}{\theta}$ if, and only if, $Y \stackrel{d}{=} SU$ where the symbol $\stackrel{d}{=}$ means equality in distribution.

Proof The assumption that the random variable S is independent of the random variable U, following the standard uniform distribution, implies that the characteristic function of the random variable SU is

$$\varphi_{SU}(u) = \int_0^1 \varphi_S(uv)dv.$$

If we use the characteristic function

$$\varphi_Y(u) = \frac{1}{i\theta u} \log \varphi_S(u)$$

and the characteristic function

$$\varphi_{SU}(u) = \int_0^1 \varphi_S(uv)dv$$

in the relationship $Y \stackrel{d}{=} SU$ then we get the integral equation

$$\frac{1}{i\theta u}\log \varphi_S(u) = \int_0^1 \varphi_S(uv)dv. \tag{3.3.10}$$

From (3.3.10) it follows the integral equation

$$\frac{1}{i\vartheta u}\log \varphi_S(u) = \frac{1}{u}\int_0^u \varphi_S(y)dy$$

or equivalently the integral equation

$$\log \varphi_S(u) = i\theta \int_0^u \varphi_S(y)dy. \tag{3.3.11}$$

Moreover (3.3.11) implies that

$$\frac{d}{du}\log \varphi_S(u) = i\theta \frac{d}{du}\int_0^u \varphi_S(y)dy. \tag{3.3.12}$$

From (3.3.12) we get the differential equation

$$\frac{d\varphi_S(u)}{du} = i\vartheta\varphi_S^2(u) \tag{3.3.13}$$

which satisfies the condition $\varphi_S(0) = 1$.

Hence (3.3.13) implies that

$$\int_0^u \frac{d\varphi_S(y)}{\varphi_S^2(y)} = \int_0^u i\vartheta dy. \tag{3.3.14}$$

From (3.3.14) we get that

$$\frac{1}{\varphi_S(u)} = 1 - i\theta u$$

or equivalently we get that

$$\varphi_S(u) = \frac{1}{1 - i\theta u}. \tag{3.3.15}$$

If $\theta = \frac{1}{\mu}$ then (3.3.15) has the form

$$\varphi_S(u) = \frac{\mu}{\mu - iu}. \tag{3.3.16}$$

Conversely, we suppose that (3.3.16) is valid. Since the random variable S with characteristic function

$$\varphi_S(u) = \frac{\mu}{\mu - iu}$$

is infinitely divisible with mean $\frac{1}{\mu}$ then

$$\varphi_Y(u) = \frac{\mu}{iu} \log \frac{\mu}{\mu - iu} \tag{3.3.17}$$

is the characteristic function of a random variable Y with probability density function $f_Y(y)$ which has a unique mode at the point 0. From (3.3.17) we get that

$$\frac{\mu}{iu} \log\left(\frac{\mu}{\mu - iu}\right) = \int_0^1 \frac{\mu}{\mu - iuv} dv. \tag{3.3.18}$$

Let U be a random variable which follows the uniform distribution with probability density function $f_U(v) = 1, 0 < v < 1$ and independent of the random variable S.

If we use the random variable Y and the random variable SU in (3.3.18) then we get that $Y \stackrel{d}{=} SU$.

The second special case of the stochastic multiplicative model $Y = SU$ arises if the random variable U follows the beta distribution with probability density function

$$f_U(v) = \nu(1 - v)^{\nu - 1}, \quad 0 < v < 1, \quad \nu > 0. \tag{3.3.19}$$

From (3.3.1), (3.3.2), (3.3.3) and (3.3.19) we get the formula

$$F_Y(y) = \nu \int_0^1 F_S\left(\frac{y}{\upsilon}\right)(1-\upsilon)^{\nu-1}d\upsilon$$

for the distribution function, the formula

$$f_Y(y) = \nu \int_0^1 \frac{1}{\upsilon} f_S\left(\frac{y}{\upsilon}\right)(1-\upsilon)^{\nu-1}d\upsilon$$

for the probability distribution function, and the formula

$$\varphi_Y(u) = \nu \int_0^1 \varphi_S(u\upsilon)(1-\upsilon)^{\nu-1}d\upsilon \tag{3.3.20}$$

for the characteristic function of the stochastic multiplicative model $Y = SU$.

The above formulas correspond to the class of ν-unimodal probability distributions.

If $\nu > 1$ then the random variable U has probability density function

$$f_U(\upsilon) = \nu(1-\upsilon)^{\nu-1}, \quad 0<\upsilon<1,$$

which is unimodal at the point 0. If $\nu > 1$ then the result of Medgyessy implies that the probability density function

$$f_Y(y) = \nu \int_0^1 \frac{1}{\upsilon} f_S\left(\frac{y}{\upsilon}\right)(1-\upsilon)^{\nu-1}d\upsilon$$

of the stochastic model $Y = SU$ has a unique mode at the point 0. For $\nu = 1$ then it easily follows the formula

$$F_Y(y) = \int_0^1 F_S\left(\frac{y}{\upsilon}\right)d\upsilon$$

for the distribution function, the formula

$$f_Y(y) = \int_0^1 \frac{1}{\upsilon} f_S\left(\frac{y}{\upsilon}\right)d\upsilon$$

for the probability density function and the formula

$$\varphi_Y(u) = \int_0^1 \varphi_S(uv)dv$$

for the characteristic function of the stochastic model $Y = SU$.

If $v = 1$ the result of Khintchine implies that the probability density function

$$f_Y(y) = \int_0^1 \frac{1}{v} f_S\left(\frac{y}{v}\right) dv$$

of the stochastic model $Y = SU$ has a unique mode at the point 0.

An interesting special case of the characteristic function (3.3.20) is the following. If the random variable S follows the gamma distribution with characteristic function

$$\varphi_S(u) = \left(\frac{\mu}{\mu - iu}\right)^3$$

and $v = 2$ then the characteristic function (3.3.20) has the form

$$\varphi_Y(u) = 2\int_0^1 \left(\frac{\mu}{\mu - iuv}\right)^3 (1 - v)dv. \tag{3.3.21}$$

From (3.3.21) it follows that

$$\varphi_Y(u) = \frac{\mu}{iu}\int_0^1 (1 - v)d\left(\frac{\mu}{\mu - iuv}\right)^2. \tag{3.3.22}$$

A factorial integration of (3.3.22) implies that

$$\varphi_Y(u) = -\frac{\mu}{iu} + \frac{\mu}{iu}\int_0^1 \left(\frac{\mu}{\mu - iuv}\right)^2 dv. \tag{3.3.23}$$

Integrating (3.3.23) we get that

$$\varphi_Y(u) = -\frac{\mu}{iu} + \frac{\mu}{iu}\left(\frac{\mu}{\mu - iu}\right)$$

or equivalently we get that

$$\varphi_Y(u) = \frac{\mu}{\mu - iu}.$$

In this case the random variable Y follows the exponential distribution with parameter μ. □

3.4 Operations of Deleting Risk Occurrences with Constant Probability

We consider a risk with frequency represented by the discrete random variable N taking values in the set $\mathbf{N}_0 = \{0, 1, 2, \ldots\}$, having probability function

$$P(N = n) = p_n, \quad n = 0, 1, 2, \ldots \tag{3.4.1}$$

and probability generating function $P_N(z)$.

An operation of reducing the frequency N of a risk is a process which is applied to the conditions and the cause of a risk. The consequence of that process is the representation of the risk frequency with a discrete random variable Y taking values in the set $\mathbf{N}_0 = \{0, 1, 2, \ldots\}$ and satisfying the relationship $P(Y < N) = 1$.

Since the random variable Y denotes the frequency of risk after applying a risk frequency reduction operation then this random variable is a function of the random variable N which denotes the frequency of risk before applying the risk frequency reduction operation. The formulation of the function between the random variable Y and the random variable N is called stochastic modeling of the risk frequency reduction operation. The result of stochastic modeling of the risk frequency reduction operation is the stochastic model denoting the risk frequency Y after applying the risk frequency reduction operation. This model is called a stochastic model of a risk frequency reduction operation. The purpose of the present section is the formulation of a random sum for the description and investigation of a risk frequency reduction operation. We suppose that in the risk with frequency N is applied the following operation of reducing the frequency N.

According to this operation each risk occurrence is retained with probability w and deleted with probability $1 - w$.

The retention–deletion of a risk occurrence is independent of the retention–deletion of any other risk occurrence. We consider the sequence of independent and identically distributed random variables $\{L_n, n = 1, 2, \ldots\}$.

The random variables of the above sequence are equally distributed with the random variable L which follows the Bernoulli distribution with probability function

$$P(L=l)=\begin{cases} w, & l=1 \\ 1-w, & l=0 \end{cases}$$

The random variable N is independent of the sequence $\{L_n, n=1,2,\ldots\}$.

If the random variable Y denotes the risk frequency after applying the risk frequency reduction operation then we get that

$$Y=L_1+L_2+\cdots+L_N. \tag{3.4.2}$$

The probability generating function $P_Y(z)$ of the random sum

$$Y=L_1+L_2+\cdots+L_N$$

is required for investigating the impact of the risk frequency reduction operation.

The following result concentrates on the evaluation of the probability generating function $P_Y(z)$.

Theorem 3.4.1 *Let N be a discrete random variable with values in the set $\mathbf{N}_0=\{0,1,2,\ldots\}$, probability function*

$$P(N=n)=p_n, \quad n=0,1,2,\ldots$$

and probability generating function $P_N(z)$.
Let $\{L_n, n=1,2,\ldots\}$ be a sequence of independent and identically distributed random variables. The random variables of the sequence $\{L_n, n=1,2,\ldots\}$ are equally distributed with the random variable L which follows the Bernoulli distribution with probability function

$$P(L=l)=\begin{cases} w, & l=1 \\ 1-w, & l=0 \end{cases}.$$

We suppose that the random variable N is independent of the sequence $\{L_n, n=1,2,\ldots\}$ and we set

$$Y=L_1+L_2+\cdots+L_N.$$

The probability generating function of the random sum

$$Y=L_1+L_2+\cdots+L_N$$

is

$$P_Y(z)=P_N(1-w+wz).$$

Proof The probability generating function of the random variable L is

$$P_L(z) = \sum_{l=0}^{1} w^l (1-w)^{1-l} z^l$$

or equivalently

$$P_L(z) = 1 - w + wz. \tag{3.4.3}$$

The evaluation of the probability generating function $P_Y(z)$ is implemented as follows. We have

$$P_Y(z) = E(z^Y)$$

or equivalently

$$P_Y(z) = E(E(z^Y|N)). \tag{3.4.4}$$

From (3.4.2) and (3.4.4) it follows that

$$P_Y(z) = \sum_{n=0}^{\infty} E(z^{L_1+\cdots+L_N}|N=n)P(N=n)$$

or equivalently

$$P_Y(z) = \sum_{n=0}^{\infty} E(z^{L_1+\cdots+L_n}|N=n)P(N=n). \tag{3.4.5}$$

Moreover, (3.4.5) implies that

$$P_Y(z) = \sum_{n=0}^{\infty} E(z^{L_1}\ldots z^{L_n}|N=n)P(N=n). \tag{3.4.6}$$

The independence of the random variable N and sequence of random variables $\{L_n, n = 1, 2, \ldots\}$ means the independence of the random variables $N, L_1, \ldots, L_n$.

Hence we get the independence of the random variables $N, z^{L_1}\ldots z^{L_n}$ and the independence of the random variables $z^{L_1}, \ldots, z^{L_n}$.

It is easily seen that (3.4.6) has the form

$$P_Y(z) = \sum_{n=0}^{\infty} E(z^{L_1})\ldots E(z^{L_n})P(N=n). \tag{3.4.7}$$

Since

$$P_L(z) = E\left(z^{L_n}\right), \quad n = 1, 2, \ldots$$

then (3.4.3) implies that

$$E\left(z^{L_n}\right) = 1 - w + wz. \tag{3.4.8}$$

From (3.4.1), (3.4.7) and (3.4.8) it follows that

$$P_Y(z) = \sum_{n=0}^{\infty} (1 - w + wz)^n p_n. \tag{3.4.9}$$

Hence (3.4.9) implies that the probability generating function of the random sum

$$Y = L_1 + L_2 + \cdots + L_N$$

has the form

$$P_Y(z) = P_N(1 - w + wz). \tag{3.4.10}$$

Some special cases of (3.4.10) are the following. We suppose that the random variable N follows the Poisson distribution with probability generating function

$$P_N(z) = e^{\lambda(z-1)}. \tag{3.4.11}$$

From (3.4.10) and (3.4.11) we get that

$$P_Y(z) = e^{\lambda(1-w+wz-1)}$$

or equivalently we get that

$$P_Y(z) = e^{\lambda w(z-1)}.$$

Hence the random variable Y follows the Poisson distribution with parameter λw.

We suppose that the random variable N follows the stable distribution with probability generating function

$$P_N(z) = e^{-c(1-z)^{\gamma}}, \quad c > 0, \ 0 < \gamma \le 1. \tag{3.4.12}$$

From (3.4.10) and (3.4.12) it follows that

$$P_Y(z) = e^{-c[1-(1-w+wz)]^{\gamma}}$$

or equivalently

$$P_Y(z) = e^{-cw^{\gamma}(1-z)^{\gamma}}$$

Hence the random variable Y follows the stable distribution with parameters cw^{γ}, γ.

We suppose that the random variable N follows the uniform distribution with probability generating function

$$P_N(z) = \frac{1-z^n}{n(1-z)}. \tag{3.4.13}$$

From (3.4.10) and (3.4.13) we get that

$$P_Y(z) = \frac{1-(1-w+wz)^n}{n[1-(1-w+wz)]}$$

or equivalently we get that

$$P_Y(z) = \frac{1-(1-w+wz)^n}{nw(1-z)}.$$

Hence the random variable Y follows the renewal distribution corresponding to the random variable Δ which follows the binomial distribution with probability generating function

$$P_{\Delta}(z) = (1-w+wz)^n.$$

We suppose that the random variable N follows the renewal distribution corresponding to the random variable H which follows the Poisson distribution with probability generating function

$$P_H(z) = e^{\lambda(z-1)}$$

The probability generating function of the random variable N has the form

$$P_N(z) = \frac{1-e^{\lambda(z-1)}}{\lambda(1-z)}. \tag{3.4.14}$$

From (3.4.10) and (3.4.14) we get that

$$P_Y(z) = \frac{1 - e^{\lambda(1-w+wz-1)}}{\lambda[1 - (1 - w + wz)]}$$

or equivalently we get that

$$P_Y(z) = \frac{1 - e^{\lambda w(z-1)}}{\lambda w(1 - z)}.$$

Hence the random variable Y follows the renewal distribution corresponding to the random variable I which follows the Poisson distribution with probability generating function

$$P_{\text{I}}(z) = e^{\lambda w(z-1)}.$$

The use of the random sum

$$Y = L_1 + L_2 + \cdots + L_N$$

as a stochastic model of a risk frequency reduction operation supports the role of random sums in describing and investigating fundamental risk management operations. □

3.5 Operations of Deleting Risk Occurrences with Random Probability

Let N be a discrete random variable with values in the set $\mathbf{N}_0 = \{0, 1, 2, \ldots\}$ and probability generating function $P_N(z)$.

We suppose that the random variable N denotes the frequency of a risk and a risk frequency reduction operation is applied to that risk. According to this risk frequency reduction operation a risk occurrence is retained with probability W or deleted with probability $1 - W$ where W is a continuous and positive random variable with probability density function $f_W(w), 0 < w < 1$.

The deletion–retention of a risk occurrence is independent of the deletion–retention of any other risk occurrence.

Let V be a discrete random variable denoting the risk frequency after applying the risk frequency reduction operation. It is obvious that the random variable V takes values in the set $\mathbf{N}_0 = \{0, 1, 2, \ldots\}$, and the random variable N with the random variable V satisfy the relationship $P(V < N) = 1$.

The purpose of the present section is the evaluation of the probability generating function $P_V(z)$ of the random variable V which denotes the frequency of risk after applying the risk frequency reduction operation.

Theorem 3.5.1 *Let N be a discrete random variable with values in the set* $\mathbf{N}_0 = \{0, 1, 2, \ldots\}$ *and probability generating function* $P_N(z)$.

We suppose that the random variable N denotes the risk frequency and a risk frequency reduction operation is applied to that risk. According to this risk frequency reduction operation a risk occurrence is retained with probability W or deleted with probability $1 - W$ where W is a continuous, positive random variable independent of the frequency of risk with probability density function $f_W(w), 0 < w < 1$.

The retention–deletion of a risk occurrence is independent of the retention–deletion of any other risk occurrence.

Let V be a discrete random variable denoting the frequency of risk after applying the risk frequency reduction operation. The random variable V takes values in the set $\mathbf{N}_0 = \{0, 1, 2, \ldots\}$ *and*

$$P_V(z) = \int_0^1 P_N(1 - w + wz) f_W(w) dw$$

is the probability generating function of the random variable V.

Proof We suppose that $W = w$.

In this case the random variable $V|W = w$ denotes the frequency of risk after applying the above risk frequency reduction operation which retains a risk occurrence with probability w.

From Sect. 3.4 it follows that the random variable $V|W = w$ is equally distributed with the random variable

$$Y = L_1 + L_2 + \cdots + L_N$$

where $\{L_n,\ n = 1, 2, \ldots\}$ is a sequence of independent random variables equally distributed with the random variable L which follows the Bernoulli distribution with probability function

$$P(L = l) = \begin{cases} w, & l = 1 \\ 1 - w, & l = 0. \end{cases}$$

The sequence of random variables $\{L_n, n = 1, 2, \ldots\}$ is independent of the random variable N.

For the probability generating function $P_V(z)$ of the random variable V we get

$$P_V(z) = E(z^V)$$

or equivalently we get

$$P_V(z) = E\big(E(z^V|W)\big). \tag{3.5.1}$$

From (3.5.1) we get

$$P_V(z) = \int_0^1 E(z^V|W = w)f_W(w)dw. \tag{3.5.2}$$

Since the random variable $V|W = w$ has the same distribution with the random sum

$$Y = L_1 + L_2 + \cdots + L_N,$$

then we get

$$E(z^V|W = w) = E(z^Y)$$

or equivalently we get

$$E(z^V|W = w) = P_Y(z). \tag{3.5.3}$$

From (3.4.10) we get

$$P_Y(z) = P_N(1 - w + wz). \tag{3.5.4}$$

Hence (3.5.2), (3.5.3) and (3.5.4) imply that the probability generating function $P_V(z)$ of the random variable V has the form

$$P_V(z) = \int_0^1 P_N(1 - w + wz)f_W(w)dw. \tag{3.5.5}$$

A special case of the probability generating function (3.5.5) is the following. We suppose that the random variable W follows the beta distribution with probability density function

$$f_W(w) = aw^{a-1}, \quad 0<w<1, \ a > 0. \tag{3.5.6}$$

From (3.5.5) and (3.5.6) it follows that the probability generating function $P_V(z)$ of the random variable V has the following form

$$P_V(z) = a\int_0^1 P_N(1 - w + wz)w^{a-1}dw. \tag{3.5.7}$$

If we set

$$y = 1 - w + wz$$

then the probability generating function (3.5.7) has the form

$$P_V(z) = \frac{a}{(1-z)^a} \int_z^1 P_N(y)(1-y)^{a-1} dy. \tag{3.5.8}$$

The discrete random variable V with values in the set $\mathbf{N}_0 = \{0, 1, 2, \ldots\}$ and probability generating function of the form (3.5.8) is called α-monotone. Some examples of α-monotone variables are the following.

We suppose that the random variable N follows the Bernoulli distribution with probability generating function

$$P_N(z) = pz + q \tag{3.5.9}$$

and probability function

$$P(N = n) = \begin{cases} p, & n = 1 \\ q, & n = 0 \end{cases}.$$

From (3.5.8) and (3.5.9) it follows that the probability generating function of the random variable V has the form

$$P_V(z) = \frac{a}{(1-z)^a} \int_z^1 (py + q)(1-y)^{a-1} dy$$

or equivalently the form

$$P_V(z) = \frac{ap}{a+1} z + \frac{aq+1}{a+1},$$

which belongs to the Bernoulli distribution with probability function

$$P(V = v) = \begin{cases} \frac{ap}{a+1}, & v = 1 \\ \frac{aq+1}{a+1}, & v = 0. \end{cases}$$

We suppose that the random variable N follows the Sibuya distribution with probability generating function

$$P_N(z) = 1 - (1-z)^\gamma, \quad 0<\gamma\leq 1. \tag{3.5.10}$$

From (3.5.8) and (3.5.10) it follows that the probability generating function of the random variable V has the form

$$P_V(z) = \frac{a}{(1-z)^a}\int_z^1 [1-(1-y)^\gamma](1-y)^{a-1}dy$$

or equivalently the form

$$P_V(z) = \frac{\gamma}{a+\gamma} + \frac{a}{a+\gamma}[1-(1-z)^\gamma]$$

which is a mixture of the probability generating function $P_{\mathrm{I}}(z) = 1$ of the random variable I with probability function $P(\mathrm{I}=0) = 1$ and the probability generating function

$$P_J(z) = 1-(1-z)^\gamma$$

of the random variable J which follows the Sibuya distribution with parameter γ.

We suppose that the random variable N follows the negative binomial distribution with probability generating function

$$P_N(z) = \left(\frac{p}{1-qz}\right)^{a+1}. \tag{3.5.11}$$

From (3.5.8) and (3.5.11) it follows that the probability generating function of the random variable V has the form

$$P_V(z) = \frac{a}{(1-z)^a}\int_z^1 \left(\frac{p}{1-qy}\right)^{a+1}(1-y)^{a-1}dy$$

or equivalently the form

$$P_V(z) = \left(\frac{p}{1-qz}\right)^a$$

which follows the negative binomial distribution with parameters p, a.

We suppose that the random variable N follows the Poisson distribution with probability generating function

$$P_N(z) = e^{\lambda(z-1)}. \tag{3.5.12}$$

From (3.5.8) with $a = 1$ and (3.5.12) it follows that the probability generating function of the random variable V has the form

$$P_V(z) = \frac{1}{1-z} \int_z^1 e^{\lambda(y-1)} dy$$

or equivalently the form

$$P_V(z) = \frac{1 - e^{\lambda(z-1)}}{\lambda(1-z)}$$

which is the probability generating function of the renewal distribution corresponding to the random variable Π which follows the Poisson distribution with probability generating function

$$P_\Pi(z) = e^{\lambda(z-1)}.$$

The class of α-monotone random variables for $a = 1$ has particular theoretical and practical significance since this class is related to the class of integral part models. These models have important applications in the description and analysis of operations for treating risks with at least one occurrence into a given time interval. □

3.6 Integral Part Models of Risk Frequency Reduction Operations

Let N be a discrete random variable with values in the set $\mathbf{N} = \{1, 2, \ldots\}$, probability function

$$P(N = n) = p_n, \quad n = 1, 2 \tag{3.6.1}$$

and probability generating function $P_N(z)$.

We consider the random variable U which follows the uniform distribution with probability density function

$$f_U(v) = 1, \quad 0 < v < 1. \tag{3.6.2}$$

We suppose that the random variable N is independent of the random variable U and we set $Y = [UN]$ where $[UN]$ denotes the integral part of the product UN.

The random variable $Y = [UN]$ is a stochastic model of particular theoretical and practical significance.

Since $P(Y < N) = 1$ then the stochastic model of integral part $Y = [UN]$ can be interpreted in the area of risk frequency reduction operations in the following way. We suppose that the random variable N denotes the frequency of a risk, then the random variable Y denotes the frequency of the risk after applying a risk frequency reduction operation. The purpose of the present section is the evaluation of the probability function

$$P(Y = y) = q_y, \quad y = 0, 1, 2\ldots,$$

the probability generating function $P_Y(z)$, and the establishment of properties and applications in the area of risk frequency reduction operations of the stochastic model $Y = [UN]$.

The following result concentrates on the evaluation of the probability function

$$P(Y = y) = q_y, \quad y = 0, 1, 2, \ldots$$

and the evaluation of the probability generating function $P_Y(z)$ of the stochastic model $Y = [UN]$.

Theorem 3.6.1 *Let N be a discrete random variable with values in the set $\mathbf{N} = \{1, 2, \ldots\}$, probability function*

$$P(N = n) = p_n, \quad n = 0, 1, 2, \ldots$$

and probability generating function $P_N(z)$.

We consider the random variable U which follows the uniform distribution with probability density function

$$f_U(v) = 1, \quad 0 < v < 1. \tag{3.6.3}$$

We suppose that the random variable N is independent of the random variable U and we set $Y = [UN]$ where $[UN]$ is the integral part of the product UN.

The stochastic model $Y = [UN]$ has probability function

$$P(Y = y) = q_y = \sum_{n=y+1}^{\infty} \frac{p_n}{n}$$

and probability generating function

$$P_Y(z) = \frac{1}{1-z} \int_z^1 \frac{P_N(y)}{y} dy.$$

Proof The evaluation of the probability function

$$P(Y = y) = q_y, \quad y = 0, 1, 2, \ldots$$

of the random variable $Y = [UN]$ is implemented by the following way. We have

$$P(Y = y) = P([UN] = y)$$

or equivalently we have

$$P(Y = y) = \sum_{n=y+1}^{\infty} P([UN] = y|N = n)P(N = n). \tag{3.6.4}$$

From (3.6.1) and (3.6.4) it follows that

$$P(Y = y) = \sum_{n=y+1}^{\infty} P([UN] = y|N = n)p_n. \tag{3.6.5}$$

Since the event $([UN] = y|N = n)$ is equivalent to the event $(y \leq UN < y + 1|N = n)$ then (3.6.5) implies that

$$P(Y = y) = \sum_{n=y+1}^{\infty} P(y \leq UN < y + 1|N = n)p_n. \tag{3.6.6}$$

Hence (3.6.6) implies that

$$P(Y = y) = \sum_{n=y+1}^{\infty} P\left(\frac{y}{N} \leq U < \frac{y+1}{N}|N = n\right)p_n$$

and

$$P(Y = y) = \sum_{n=y+1}^{\infty} P\left(\frac{y}{n} \leq U < \frac{y+1}{n}|N = n\right)p_n. \tag{3.6.7}$$

Since the random variable U is independent of the random variable N then (3.6.7) has the form

$$P(Y = y) = \sum_{n=y+1}^{\infty} P\left(\frac{y}{n} \leq U < \frac{y+1}{n}\right)p_n. \tag{3.6.8}$$

From (3.6.2) or (3.6.3) and (3.6.8) it follows that

$$P(Y = y) = \sum_{n=y+1}^{\infty} \left(\frac{y+1}{n} - \frac{y}{n} \right) p_n.$$

Hence the probability function of the random variable $Y = [UN]$ is

$$P(Y = y) = q_y = \sum_{n=y+1}^{\infty} \frac{p_n}{n}, \quad y = 0, 1, 2, \ldots \tag{3.6.9}$$

From (3.6.9) it follows that

$$P(Y = y) = q_y > P(Y = y+1) = q_{y+1}, \quad y = 0, 1, 2\ldots \tag{3.6.10}$$

and (3.6.10) implies that the probability function

$$P(Y = y) = q_y, \quad y = 0, 1, 2\ldots$$

of the random variable $Y = [UN]$ has a unique mode at the point 0. Hence risk avoidance is the most probable event after applying the risk frequency reduction operation.

The evaluation of the probability generating function $P_Y(z)$ of the random variable $Y = [UN]$ is implemented by the following way. We have

$$P_Y(z) = E(z^Y)$$

or equivalently we have

$$P_Y(z) = \sum_{y=0}^{\infty} z^y P(Y = y). \tag{3.6.11}$$

From (3.6.9) and (3.6.11) it follows that

$$P_Y(z) = \sum_{y=0}^{\infty} z^y \left(\sum_{n=y+1}^{\infty} \frac{p_n}{n} \right). \tag{3.6.12}$$

Hence (3.6.12) implies that

$$P_Y(z) = \sum_{n=1}^{\infty} \frac{p_n}{n} \left(\sum_{y=0}^{n-1} z^y \right). \tag{3.6.13}$$

From (3.6.13) it follows that

$$P_Y(z) = \sum_{n=1}^{\infty} \frac{p_n}{n}\left(1 + z + \cdots + z^{n-1}\right). \tag{3.6.14}$$

Hence (3.6.14) implies that

$$P_Y(z) = \sum_{n=1}^{\infty} \frac{p_n}{n} \cdot \frac{1 - z^n}{1 - z}.$$

It is easily seen that

$$P_Y(z) = \sum_{n=1}^{\infty} p_n \frac{1}{1 - z} \int_z^1 w^{n-1} dw. \tag{3.6.15}$$

From (3.6.15) we get that

$$P_Y(z) = \frac{1}{1 - z} \sum_{n=1}^{\infty} p_n \int_z^1 \frac{w^n}{w} dw. \tag{3.6.16}$$

Moreover (3.6.16) implies

$$P_Y(z) = \frac{1}{1 - z} \int_z^1 \left(\frac{1}{w} \sum_{n=1}^{\infty} p_n w^n \right) dw. \tag{3.6.17}$$

Since

$$P_N(z) = \sum_{n=1}^{\infty} p_n z^n$$

is the probability generating function of the random variable N then (3.6.17) implies that the probability generating function of the random variable $Y = [UN]$ is

$$P_Y(z) = \frac{1}{1 - z} \int_z^1 \frac{P_N(w)}{w} dw. \tag{3.6.18}$$

The consideration of special cases of the probability distribution of the stochastic model of integral part $Y = [UN]$ is of particular practical interest. The use of the probability generating function (3.6.18) is important in considering such cases. Some special cases of (3.6.18) are the following.

We suppose that the random variable N, denoting the frequency of risk before applying the risk frequency reduction operation, follows the geometric type II distribution with probability generating function

$$P_N(z) = \frac{pz}{1-qz}. \tag{3.6.19}$$

From (3.6.18) and (3.6.19) it follows that the probability generating function of the random variable $Y = [UN]$, which denotes the frequency of risk after applying the risk frequency reduction operation, has the form

$$P_Y(z) = \frac{1}{1-z}\int_z^1 \frac{1}{w}\cdot\frac{pw}{1-w}dw$$

or equivalently the form

$$P_Y(z) = \frac{p}{q}\log\left(\frac{p}{1-qz}\right)\Big/(z-1). \tag{3.6.20}$$

The probability generating function (3.6.20) belongs to the renewal distribution corresponding to the probability distribution of the random variable Δ which follows the logarithmic distribution with probability generating function

$$P_\Delta(z) = \frac{\log(1-qz)}{\log p}.$$

We suppose that the random variable N has the form $N = B + 1$ where B is a random variable following the Poisson distribution with probability generating function

$$P_B(z) = e^{\lambda(z-1)}$$

The probability generating function of the random variable N is

$$P_N(z) = ze^{\lambda(z-1)}. \tag{3.6.21}$$

From (3.6.18) and (3.6.21) it follows that the probability generating function of the random variable $Y = [UN]$ has the form

$$P_Y(z) = \frac{1}{1-z}\int_z^1 \frac{1}{w}\cdot we^{\lambda(w-1)}dw$$

or equivalently the form

$$P_Y(z) = \frac{1 - e^{\lambda(z-1)}}{\lambda(1-z)}. \tag{3.6.22}$$

The probability generating function (3.6.22) belongs to the renewal distribution corresponding to the probability distribution of the random variable B which follows the Poisson distribution with probability generating function

$$P_B(z) = e^{\lambda(z-1)}.$$

We suppose that the random variable N follows the uniform distribution with probability generating function

$$P_N(z) = \frac{z(1-z^n)}{n(1-z)}. \tag{3.6.23}$$

From (3.6.18) and (3.6.23) it follows that the probability generating function of the random variable $Y = [UN]$ has the form

$$P_Y(z) = \frac{1}{1-z} \int_z^1 \frac{1}{w} \cdot \frac{w(1-w^n)}{n(1-w)} dw$$

or equivalently the form

$$P_Y(z) = \sum_{\kappa=1}^{n} \frac{1}{n} \cdot \frac{1-z^\kappa}{\kappa(1-z)}. \tag{3.6.24}$$

The probability generating function (3.6.24) belongs to the distribution which is a uniform mixture of uniform distributions.

We suppose that the random variable N follows the degenerate distribution with probability function $P(N = n) = 1$ and probability generating function

$$P_N(z) = z^n. \tag{3.6.25}$$

From (3.6.18) and (3.6.25) it follows that the probability generating function of the random variable $Y = [UN]$ has the form

$$P_Y(z) = \frac{1}{1-z} \int_z^1 \frac{1}{w} \cdot w^n dw$$

or equivalently the form

$$P_Y(z) = \frac{1 - z^n}{n(1 - z)}. \tag{3.6.26}$$

The probability generating function (3.6.26) belongs to the uniform distribution.

We suppose that the random variable N has the form $N = H + 1$ where H is a random variable following the Sibuya distribution with probability generating function

$$P_H(z) = 1 - (1 - z)^{\gamma}, \quad 0 \leq \gamma \leq 1.$$

The probability generating function of the random variable N has the form

$$P_N(z) = z[1 - (1 - z)^{\gamma}]. \tag{3.6.27}$$

From (3.6.18) and (3.6.27) it follows that the probability generating function of the random variable $Y = [UN]$ has the form

$$P_Y(z) = \frac{1}{1 - z} \int_z^1 \frac{1}{w} \cdot w[1 - (1 - w)^{\gamma}] dw$$

or equivalently the form

$$P_Y(z) = \frac{\gamma}{\gamma + 1} + \frac{1}{\gamma + 1}[1 - (1 - z)^{\gamma}]. \tag{3.6.28}$$

The probability generating function (3.6.28) is a mixture of the probability generating function $P_{\mathrm{I}}(z) = 1$ of the random variable I with probability function $P(\mathrm{I} = 0) = 1$ and the probability generating function

$$P_J(z) = 1 - (1 - z)^{\gamma}$$

of the random variable J which follows the Sibuya distribution with parameters γ.

The following result provides an interesting interpretation of the class of discrete self-decomposable probability distributions in the area of risk frequency reduction operations. □

Theorem 3.6.2 *Let N be a discrete random variable with values in the set $\mathbf{N} = \{1, 2, \ldots\}$ and probability generating function $P_N(z)$, C a discrete random variable with values in the set $\mathbf{N}_0 = \{0, 1, 2, \ldots\}$, probability generating function $P_C(z)$ and U a random variable following the uniform distribution with probability density function $f_U(v) = 1, 0 < v < 1$.*

We suppose that the random variables N, C, U are independent. The probability generating function $P_C(z)$ of the random variable C has the form

$$P_C(z) = \exp\left(-\int_z^1 \frac{1 - P_N(w)}{1 - w} dw\right)$$

if, and only if,

$$C \stackrel{d}{=} [U(N + C + 1)]$$

where the symbol $\stackrel{d}{=}$ *means equality in distribution.*

Proof We consider the random variable

$$\Pi = [U(N + C + 1)]$$

The evaluation of the probability generating function $P_\Pi(z)$ requires the proof of independence of the random variables U and $L = N + C + 1$.

Let

$$\varphi_U(u) \tag{3.6.29}$$

be the characteristic function of the random variable U, $\varphi_L(\xi)$ the characteristic function of the random variable $L = N + C + 1$, and $\varphi_{U,L}(u, \xi)$ the characteristic function of the vector (U, L). The establishment of the relationship

$$\varphi_{U,L}(u, \xi) = \varphi_U(u)\varphi_L(\xi) \tag{3.6.30}$$

is required for the proof of the independence of the random variables U, L. The characteristic function of the random variable $L = N + C + 1$ is

$$\varphi_L(\xi) = E\left(e^{i\xi(N+C+1)}\right)$$

or equivalently

$$\varphi_L(\xi) = e^{i\xi} E\left(e^{i\xi N} \cdot e^{i\xi C}\right). \tag{3.6.31}$$

The independence of the random variables N, C, U implies the independence of the random variables N, C.

Hence the random variables $e^{i\xi N}$, $e^{i\xi C}$ are independent. Since

$$\varphi_N(\xi) = E\left(e^{i\xi N}\right) \tag{3.6.32}$$

is the characteristic function of the random variable N and

$$\varphi_C(\xi) = E\left(e^{i\xi C}\right) \tag{3.6.33}$$

is the characteristic function of the random variable C then (3.6.31) has the form

$$\varphi_L(\xi) = e^{i\xi}\varphi_{\mathrm{N}}(\xi)\varphi_C(\xi). \tag{3.6.34}$$

From (3.6.30) and (3.6.34) it follows the establishment of the relationship

$$\varphi_{U,L}(u,\xi) = \varphi_U(u)\varphi_{\mathrm{N}}(\xi)\varphi_C(\xi)e^{i\xi}$$

is required for the proof of the independence of the random variables U and $L = N + C + 1$.

We have

$$\varphi_{U,L}(u,\xi) = E\left(e^{iuU+i\xi(N+C+1)}\right).$$

Hence

$$\varphi_{U,L}(u,\xi) = E\left(E\left(e^{iuU+i\xi N+i\xi C+i\xi}|C\right)\right)$$

or equivalently

$$\varphi_{U,L}(u,\xi) = \sum_{c=0}^{\infty} E\left(e^{iuU+i\xi N+i\xi C+i\xi}|C=c\right)P(C=c). \tag{3.6.35}$$

From (3.6.35) we get that

$$\varphi_{U,L}(u,\xi) = \sum_{c=0}^{\infty} E\left(e^{iuU+i\xi N+i\xi c+i\xi}|C=c\right)P(C=c) \tag{3.6.36}$$

The independence of the random variables N, C, U implies the independence of the random variables $e^{i\xi N}$, e^{iuU}, C.

Hence (3.6.36) has the form

$$\varphi_{U,L}(v,\xi) = \sum_{y=0}^{\infty} E\left(e^{iuU}e^{i\xi N}e^{i\xi c}e^{i\xi}\right)P(C=c). \tag{3.6.37}$$

The independence of the random variables $e^{i\xi N}$, e^{iuU}, C implies the independence of the random variables $e^{i\xi N}$, e^{iuU}.

Hence (3.6.37) has the form

$$\varphi_{U,L}(\upsilon,\xi) = \sum_{y=0}^{\infty} e^{i\xi} E(e^{iuU}) E(e^{i\xi N}) E(e^{i\xi c}) P(C = c). \tag{3.6.38}$$

From (3.6.29), (3.6.32), (3.6.33) and (3.6.38) it follows that

$$\varphi_{U,L}(\upsilon,\xi) = \varphi_U(u)\varphi_{\mathrm{N}}(\xi)\varphi_C(\xi)e^{i\xi}.$$

Hence the random variables U and $L = N + C + 1$ are independent. The probability generating function of the random variable $L = N + C + 1$ is

$$P_L(z) = zP_N(z)P_C(z). \tag{3.6.39}$$

From (3.6.18) and (3.6.39) it follows that the probability generating function of the random variable $\Pi = [U(N + C + 1)]$ has the form

$$P_\Pi(z) = \frac{1}{1-z}\int_z^1 P_N(w)P_C(w)dw. \tag{3.6.40}$$

If we use the probability generating function $P_C(z)$ and the probability generating function

$$P_\Pi(z) = \frac{1}{1-z}\int_z^1 P_N(w)P_C(w)dw$$

in the relationship

$$C \stackrel{d}{=} [U(N + C + 1)]$$

then we get the integral equation

$$P_C(z) = \frac{1}{1-z}\int_z^1 P_N(w)P_C(w)dw. \tag{3.6.41}$$

From (3.6.41) it follows that

$$(1-z)P_C(z) = \int_z^1 P_N(w)P_C(w)dw. \tag{3.6.42}$$

If we differentiate (3.6.42) then we get the differential equation

$$(1-z)\frac{dP_C(z)}{dz} - P_C(z) = -P_N(z)P_C(z) \tag{3.6.43}$$

which satisfies the conditions $P_C(z) = 1$ and $P_N(z) = 1$.

If $z \neq 1$ then the differential equation (3.6.43) has the form

$$\frac{dP_C(z)}{dz} = \frac{1-P_N(z)}{1-z}P_C(z). \tag{3.6.44}$$

From (3.6.44) we get that

$$\int_z^1 \frac{dP_C(w)}{P_C(w)} = \int_z^1 \frac{1-P_N(w)}{1-w}dw. \tag{3.6.45}$$

Hence (3.6.45) implies that

$$\log P_C(z) = -\int_z^1 \frac{1-P_N(w)}{1-w}dw, \quad 0 \le z < 1,$$

or equivalently

$$P_C(z) = \exp\left(-\int_z^1 \frac{1-P_N(w)}{1-w}dw\right), \quad |z| \le 1. \tag{3.6.46}$$

Conversely, we suppose that (3.6.46) is valid. We have

$$\exp\left(-\int_z^1 \frac{1-P_N(w)}{1-w}dw\right) = \frac{1}{1-z}\int_z^1 P_N(w)\exp\left(-\int_w^1 \frac{1-P_N(y)}{1-y}dy\right)dw. \tag{3.6.47}$$

From the independence of the random variables $U, L = N + C + 1$ and (3.6.47) it follows that

$$C \stackrel{d}{=} [U(N + C + 1)].$$

The probability generating function

$$P_C(z) = \exp\left(- \int_z^1 \frac{1 - P_N(w)}{1 - w} dw \right)$$

belongs to a self-decomposable probability distribution. Since the discrete self-decomposable probability distributions are unimodal then the above result is of particular practical interest. Moreover, the probability generating function

$$P_C(z) = \exp\left(- \int_z^1 \frac{1 - P_N(w)}{1 - w} dw \right)$$

is the probability generating function of the stochastic integral part model

$$C \stackrel{d}{=} [U(N + C + 1)].$$

Hence (3.6.10) implies that the probability function $P(C = c) = q_c, c = 0, 1, 2,$ which corresponds to the probability generating function

$$P_C(z) = \exp\left(- \int_z^1 \frac{1 - P_N(w)}{1 - w} dw \right)$$

satisfies the relationship

$$P(C = c) = q_c > P(C = c + 1) = q_{c+1}$$

or equivalently the probability function $P(C = c) = q_c, c = 0, 1, 2,$ is unimodal at the point 0. Since the random variable Y denotes the frequency of risk after applying a risk frequency reduction operation then the unimodality at the point 0 of the probability function $P(C = c) = q_c, c = 0, 1, 2,$ means that risk avoidance is the most probable event after applying that risk frequency reduction operation.

The consideration of special cases of the probability generating function (3.6.46) of the stochastic integral part model

$$C \stackrel{d}{=} [U(N + C + 1)]$$

supports the role of that model in practical applications.

We suppose that the random variable N follows the Sibuya distribution with probability generating function

$$P_N(z) = 1 - (1 - z)^{\gamma}, \quad 0 < \gamma \leq 1. \tag{3.6.48}$$

From (3.6.46) and (3.6.48) we get the probability generating function

$$P_C(z) = \exp\left(-\int_z^1 \frac{1-[1-(1-w)^\gamma]}{1-w}dw\right)$$

or equivalently the probability generating function

$$P_C(z) = \exp\left[-\frac{1}{\gamma}(1-z)^\gamma\right]$$

which belongs to the class of stable probability distributions.

We suppose that the random variable N follows the geometric type II distribution with probability generating function

$$P_N(z) = \frac{pz}{1-qz}. \tag{3.6.49}$$

From (3.6.46) and (3.6.49) we get the probability generating function

$$P_C(z) = \exp\left[-\int_z^1 \frac{1-\left(\frac{pw}{1-qw}\right)}{1-w}dw\right]$$

or equivalently the probability generating function

$$P_C(z) = \left(\frac{p}{1-qz}\right)^{1/q}$$

which belongs to the negative binomial distribution with parameters $p, \frac{1}{q}$.

The interpretation of the above theoretical result in the area of risk frequency reduction operations is implemented by the following way. We consider two independent risks of the same kind. We assume that the discrete random variable N with values in the set $\mathbf{N} = \{1, 2, \ldots\}$ denotes the frequency of one risk and the random variable $C+1$ denotes the frequency of the other risk, where C is a discrete random variable with values in the set $\mathbf{N}_0 = \{0, 1, 2, \ldots\}$.

A risk frequency reduction operation is applied to these risks. The kind of risks and the way of applying the risk frequency reduction operation permit the entire consideration of these risks. We suppose that the random variable U, having probability density function $f_U(\upsilon) = 1, 0<\upsilon<1$, denotes the impact of applying the risk frequency reduction operation to the total frequency $N+C+1$ of the two risks. Hence the random variable $[U(N+C+1)]$ denotes the total frequency of the two risks after applying the risk frequency reduction operation. We suppose that the random variables N, C, U are independent. The equality in distribution of the random variables C and $[U(N+C+1)]$ implies that the probability generating function $P_C(z)$ of the random variable C has the form

$$P_C(z) = \exp\left(-\int_z^1 \frac{1-P_N(w)}{1-w}dw\right),$$

and conversely the consideration of the probability generating function

$$P_C(z) = \exp\left(-\int_z^1 \frac{1-P_N(w)}{1-w}dw\right)$$

as probability generating function of the random variable C implies the equality in distribution of the random variable C and the random variable $[U(N+C+1)]$.

The class of discrete renewal distributions and a characterization of discrete distributions having a unique mode at the point 0 will be used for establishing of a result with applications in stochastic modeling of risk frequency reduction operations.

Let X be a discrete random variable with values in the set $\mathbf{N}_0 = \{0, 1, 2, \ldots\}$, probability generating function $P_X(z)$, and finite mean d.

The discrete random variable R with values in the set $\mathbf{N}_0 = \{0, 1, 2, \ldots\}$, and following the renewal distribution corresponding to the distribution of the random variable X, has probability generating function

$$P_R(z) = \frac{1-P_X(z)}{d(1-z)}.$$

The formula of the probability generating function of the random variable R which follows the renewal distribution corresponding to the distribution of the random variable X is valid even if the set of values of the random variable X is $\mathbf{N} = \{1, 2, \ldots\}$.

The following result has been established by Medgyessy.

Let A be a discrete random variable with values in the set $\mathbf{N}_0 = \{0, 1, 2, \ldots\}$ and probability generating function $P_A(z)$.

The probability function of the random variable A has a unique mode at the point 0 if, and only if, the probability generating function $P_A(z)$ of the random variable A has the form

$$P_A(z) = \frac{1-P_\Psi(z)}{\eta(1-z)}$$

where $P_\Psi(z)$ is the probability generating function of a uniquely defined random variable Ψ with value in the set $\mathbf{N} = \{1, 2, \ldots\}$ and $\eta = 1/P_A(0)$.

If the discrete random variable X has probability generating function $P_X(z)$, finite mean d, and values in the set $\mathbf{N} = \{1, 2, \ldots\}$ or the set $\mathbf{N}_0 = \{0, 1, 2, \ldots\}$ then the random variable Y with values in the set $\mathbf{N}_0 = \{0, 1, 2, \ldots\}$ and which follows the renewal distribution corresponding to the distribution of the random variable X has probability generating function

$$P_R(z) = \frac{1 - P_X(z)}{d(1 - z)}$$

which is a transformation of the probability generating function $P_X(z)$.

Hence the investigation of properties, preserved under this transformation, is of particular theoretical and practical interest. □

Theorem 3.6.3 *Let K be a discrete random variable with values in the set $\mathbf{N}_0 = \{0, 1, 2, \ldots\}$, probability generating function $P_K(z)$ and finite mean m, S be a discrete random variable following the renewal distribution corresponding to the distribution of the random variable K, T be a discrete random variable with values in the set $\mathbf{N}_0 = \{0, 1, 2, \ldots\}$, and probability generating function $P_T(z)$, and U be a random variable following the uniform distribution with probability density function $f_U(v) = 1, 0 < v < 1$.*

If the random variables K, T, U are independent then the probability distribution of the random variable K is self-decomposable if, and only if,

$$S \overset{d}{=} [U(T + K + 1)]$$

and the probability function of the random variable T is unimodal at the point 0, where the symbol $\overset{d}{=}$ means equality in distribution.

Proof Since the random variable S follows the renewal distribution corresponding to the distribution of the random variable K then the probability generating function of the random variable S is

$$P_S(z) = \frac{1 - P_K(z)}{m(1 - z)}. \tag{3.6.50}$$

From Theorem 3.6.2 it follows that the random variable U is independent of the random variable $T + K + 1$.

Hence (3.6.40) implies that the probability generating function of the random variable $E = [U(T + K + 1)]$ has the form

$$P_E(z) = \frac{1}{1 - z} \int_z^1 P_T(w) P_K(w) dw. \tag{3.6.51}$$

If we use the probability generating function in (3.6.50), or equivalently the probability generating function

$$P_S(z) = \frac{1 - P_K(z)}{m(1-z)},$$

and the probability generating function in (3.6.51), or equivalently the probability generating function

$$P_E(z) = \frac{1}{1-z}\int_z^1 P_T(w)P_K(w)dw$$

in the assumption of equality in distribution then we get the integral equation

$$\frac{1 - P_K(z)}{m(1-z)} = \frac{1}{1-z}\int_z^1 P_T(w)P_K(w)dw. \tag{3.6.52}$$

From (3.6.52) it follows that

$$1 - P_K(z) = m\int_z^1 P_T(w)P_K(w)dw. \tag{3.6.53}$$

If we differentiate (3.6.53) then we get the differential equation

$$\frac{dP_K(z)}{dz} = mP_T(z)P_K(z). \tag{3.6.54}$$

which satisfies the conditions $P_K(1) = 1$ and $P_T(1) = 1$.

The differential equation (3.6.54) has the form

$$\frac{dP_K(z)}{dz} = mP_T(z)P_K(z). \tag{3.6.55}$$

From (3.6.55) it follows that

$$\int_z^1 \frac{dP_K(w)}{P_K(w)} = m\int_z^1 P_T(w)dw \tag{3.6.56}$$

Hence (3.6.56) implies that

$$\log P_K(z) = -m \int_z^1 P_T(w)dw, \quad 0 \le z < 1$$

or equivalently

$$P_K(z) = \exp\left(-m \int_z^1 P_T(w)dw \right). \tag{3.6.57}$$

Since the probability function of the random variable T has a unique mode at the point 0 then the probability generating function $P_T(z)$ of the random variable T has the form

$$P_T(z) = \frac{1 - P_V(z)}{\rho(1-z)} \tag{3.6.58}$$

where $P_V(z)$ is the probability generating function of a uniquely defined random variable with values in the set $\mathbf{N} = \{1, 2, \ldots\}$ and

$$\rho = \frac{1}{P_T(0)}. \tag{3.6.59}$$

From (3.6.57), (3.6.58), and (3.6.59) it follows that the probability generating $P_K(z)$ of the random variable K has the form

$$P_K(z) = \exp\left(-\sigma \int_z^1 \frac{1 - P_V(w)}{1 - w} dw \right) \tag{3.6.60}$$

where $\sigma = m/\rho$

From (3.6.60) it follows that the distribution of the random variable K is self-decomposable. Conversely, we suppose that the distribution of the random variable K is self-decomposable. Hence the probability generating function of the random variable K has the form

$$P_K(z) = \exp\left(-\sigma \int_z^1 \frac{1 - P_V(w)}{1 - w} dw \right), \quad |z| \le 1, \tag{3.6.61}$$

where $P_V(z)$ is the probability generating function of a uniquely defined random variable V with values in the set $\mathbf{N} = \{1, 2, \ldots\}$ and $\sigma > 0$.

Since the random variable K has finite mean m then (3.6.61) implies that the random variable V has finite mean $\frac{m}{\sigma}$.

We set $\rho = \frac{m}{\sigma}$.

The random variable S, which follows the renewal distribution corresponding to the distribution of the random variable K, has probability generating function

$$P_S(z) = \frac{1 - \exp\left(-\sigma \int_z^1 \frac{1-P_V(w)}{1-w} dw\right)}{m(1-z)}. \tag{3.6.62}$$

The probability generating function (3.6.62) has the form

$$P_S(z) = \frac{1}{1-z} \int_z^1 \frac{1-P_V(w)}{\rho(1-w)} \exp\left(-\sigma \int_w^1 \frac{1-P_V(y)}{1-y} dy\right) dw \tag{3.6.63}$$

where

$$P_T(z) = \frac{1-P_V(z)}{\rho(1-z)} \tag{3.6.64}$$

is the probability generating function of the random variable T with probability function having a unique mode at the point 0. Since the random variable U is independent of the random variable $T+K+1$ then (3.6.18) implies that the probability generating function of the random variable $E = [U(T+K+1)]$ is

$$P_E(z) = \frac{1}{1-z} \int_z^1 P_T(w) \exp\left(-\sigma \int_w^1 \frac{1-P_V(y)}{1-y} dy\right) dw. \tag{3.6.65}$$

From (3.6.63), (3.6.64) and (3.6.65) it follows that

$$S \stackrel{d}{=} [U(T+K+1)].$$

The interpretation of the above theoretical result in the area of risk reduction operations is implemented by the following way. We consider two independent risks of the same kind. We suppose that the discrete random variable T with values in the set $\mathbf{N}_0 = \{0, 1, 2, \ldots\}$ denotes the frequency of a risk and the random variable $K+1$ denotes the frequency of another risk, where K is a discrete random variable with values in the set $\mathbf{N}_0 = \{0, 1, 2, \ldots\}$ and finite mean. A risk frequency reduction operation is applied to these risks. The kind of risks and the way of applying the risk frequency reduction operation permit the entire consideration of these risks. We suppose that the random variable U with probability density function

$$f_U(\upsilon) = 1, \quad 0<\upsilon<1$$

represents the impact of the risk frequency reduction operation on the total frequency $T+K+1$ of the two risks. Hence the random variable $[U(T+K+1)]$ denotes the total frequency of the two risks after applying the risk frequency reduction operation. We suppose that the random variables K, T, U are independent. The random variable S follows the renewal distribution corresponding to the distribution of the random variable K.

The unimodality at the point 0 of the probability function of the random variable T and the equality in distribution of the random variables S and $[U(T+K+1)]$, imply that the distribution of the random variable K is self-decomposable. Conversely, the selfdecomposability of the distribution of the random variable K implies the equality in distribution of the random variables S and $[U(T+K+1)]$ with the random variable T having probability function which is unimodal at the point 0.

The contribution of the above result consists of establishing a fundamental relationship between the class of discrete renewal distributions and the class of discrete self-decomposable distributions. The establishment of that relationship is based on the use of a stochastic integral part model. That implies the consideration of the class of discrete renewal distributions and the class of discrete self-decomposable distributions as strong analytical tools of stochastic modeling activities in the area of risk frequency reduction operations. □

3.7 Cost of an Operation of Deleting Risk Occurrences with Constant Probability

We consider an organization facing a risk at the time point 0. We suppose that $\{X_n, n=1,2,\ldots\}$ is a sequence of continuous, positive, independent, and identically distributed random variables. The random variables of the sequence are equally distributed with the random variable X which has characteristic function

$$\varphi_X(u). \tag{3.7.1}$$

The random variable $X_n, n=1,2,\ldots$ denotes the time between the $(n-1)$th and the nth risk occurrence. A risk frequency reduction operation is applied to the risk. According to this operation, a risk occurrence is retained with probability p and deleted with probability $1-p$.

The retention–deletion of a risk occurrence is independent of the retention–deletion of any other risk occurrence. Let N be a random variable denoting the number of risk occurrences required to get the first retained risk occurrence. The random variable N follows the geometric type II distribution with probability function

$$P(N = n) = pq^{n-1}, \quad n = 1, 2, \ldots \tag{3.7.2}$$

and probability generating function

$$P_N(z) = \frac{pz}{1 - qz}. \tag{3.7.3}$$

The sequence of continuous, positive, independent, and identically distributed random variables $\{X_n, n = 1, 2, \ldots\}$ is independent of the random variable N.

The random sum

$$T = X_1 + X_2 + \cdots + X_N$$

denotes the time of the first retained risk occurrence. From (2.8.4), (3.7.1) and (3.7.3) it follows that the characteristic function of the random sum

$$T = X_1 + X_2 + \cdots + X_N$$

is

$$\varphi_T(u) = \frac{p\varphi_X(u)}{1 - q\varphi_X(u)}. \tag{3.7.4}$$

Let W be a continuous and positive random variable with probability density function $f_W(w), w > 0$.

We suppose that the random variable W denotes the cost per unit of time for applying the risk frequency reduction operation. Hence the random variable

$$V = (X_1 + X_2 + \cdots + X_N)W$$

denotes the cost for applying the risk frequency reduction operation up to the time of the first retained risk occurrence. The purpose of the present section is the establishment of properties and applications in the discipline of risk management of the stochastic model

$$V = (X_1 + X_2 + \cdots + X_N)W.$$

The following result establishes sufficient conditions for the evaluation of the characteristic function of the stochastic model

$$V = (X_1 + X_2 + \cdots + X_N)W.$$

Theorem 3.7.1 *Let N be a discrete random variable with values in the set $\mathbf{N} = \{1, 2, \ldots\}$ following the geometric type II distribution with probability function*

$$P(N = n) = pq^{n-1}, \quad n = 1, 2, \ldots \tag{3.7.5}$$

and probability generating function

$$P_N(z) = \frac{pz}{1 - qz}. \tag{3.7.6}$$

We suppose that $\{X_n, \ n = 1, 2, \ldots\}$ *is a sequence of continuous, positive, and independent random variables. The random variables of the sequence are equally distributed with the random variable* X *having characteristic function*

$$\varphi_X(u) \tag{3.7.7}$$

and we set

$$T = X_1 + X_2 + \cdots + X_N.$$

Let W *be a continuous and positive random variable with probability density function* $f_W(w), w > 0$ *and we set*

$$V = (X_1 + X_2 + \cdots + X_N)W. \tag{3.7.8}$$

If N, W *and* $\{X_n, n = 1, 2, \ldots\}$ *are independent then the characteristic function of the stochastic model*

$$V = (X_1 + X_2 + \cdots + X_N)W$$

is

$$\varphi_V(u) = \int_0^\infty \frac{p\varphi_X(uw)}{1 - q\varphi_X(uw)} f_W(w)dw.$$

Proof From the assumption that N, W and $\{X_n, n = 1, 2, \ldots\}$ are independent, we get the independence of N and $\{X_n, n = 1, 2, \ldots\}$.

From (2.8.4), (3.7.6) and (3.7.7) it follows that the characteristic function of the random sum

$$T = X_1 + X_2 + \cdots + X_N$$

is

$$\varphi_T(u) = \frac{p\varphi_X(u)}{1 - q\varphi_X(u)}.$$

Let $\varphi_W(\xi)$ be the characteristic function of the random variable W and $\varphi_{T,W}(u,\xi)$ the characteristic function of the vector (T,W).

We shall prove that the assumption of independence of N, W and $\{X_n, n = 1,2,\ldots\}$ implies the relationship

$$\varphi_{T,W}(u,\xi) = \frac{p\varphi_X(u)}{1 - q\varphi_X(u)}\varphi_W(\xi)$$

or equivalently the relationship

$$\varphi_{T,W}(u,\xi) = \varphi_T(u)\varphi_W(\xi). \tag{3.7.9}$$

The establishment of the relationship (3.7.9) implies the independence of the random sum

$$T = X_1 + X_2 + \cdots + X_N$$

and the random variable W.

The independence of the random variables

$$T = X_1 + X_2 + \cdots + X_N$$

and W makes possible the evaluation of the characteristic function of the stochastic model

$$V = (X_1 + X_2 + \cdots + X_N)W$$

from the probability generating function

$$P_N(z) = \frac{pz}{1 - qz}$$

of the random variable N, the characteristic function $\varphi_X(u)$ of the random variable X and the probability density function $f_W(w)$ of the random variable W.

The proof of the relationship $\varphi_{T,W}(u,\xi) = \varphi_T(u)\varphi_W(\xi)$ is implemented by the following way.

We have

$$\varphi_{T,W}(u,\xi) = E\left(e^{iuT+i\xi W}\right)$$

or equivalently

$$\varphi_{T,W}(u,\xi) = E\left(E\left(e^{iuT+i\xi W}|N\right)\right). \tag{3.7.10}$$

From (3.7.10) it follows that

$$\varphi_{T,W}(u,\xi) = \sum_{n=1}^{\infty} E\left(e^{iu(X_1+\cdots+X_N)+i\xi W}|N=n\right)P(N=n)$$

or equivalently

$$\varphi_{T,W}(u,\xi) = \sum_{n=1}^{\infty} E\left(e^{iuX_1+\cdots+iuX_n+i\xi W}|N=n\right)P(N=n). \tag{3.7.11}$$

Moreover (3.7.5) and (3.7.11) imply that

$$\varphi_{T,W}(u,\xi) = \sum_{n=1}^{\infty} E\left(e^{iuX_1+\cdots+iuX_n+i\xi W}|N=n\right)pq^{n-1}. \tag{3.7.12}$$

From the assumption that N, W and $\{X_n, n=1,2,\ldots\}$ are independent it follows the independence of the random variables N, W, $X_1,\ldots,X_n, n=1,2,\ldots$.

The independence of the above random variables implies that the random variables N, $e^{i\xi W}, e^{iuX_1},\ldots,e^{iuX_n}$ are independent. Hence (3.7.12) implies that

$$\varphi_{T,W}(u,\xi) = \sum_{n=1}^{\infty} E\left(e^{iuX_1}\ldots e^{iuX_n}e^{i\xi W}\right)pq^{n-1}. \tag{3.7.13}$$

Moreover, the independence of the random variables N, $e^{i\xi W}, e^{iuX_1},\ldots,e^{iuX_n}, n=1,2,\ldots$ implies the independence of the random variables $e^{i\xi W}, e^{iuX_1},\ldots,e^{iuX_n}, n=1,2,\ldots$.

Hence (3.7.13) has the form

$$\varphi_{T,W}(u,\xi) = \sum_{n=1}^{\infty} E\left(e^{iuX_1}\right)\ldots E\left(e^{iuXn}\right)E\left(e^{i\xi W}\right)pq^{n-1}$$

or equivalently the form

$$\varphi_{T,W}(u,\xi) = E\left(e^{i\xi W}\right)\sum_{n=1}^{\infty} E\left(e^{iuX_1}\right)\ldots E\left(e^{iuXn}\right)pq^{n-1}. \tag{3.7.14}$$

Since $\varphi_W(\xi) = E\left(e^{i\xi W}\right)$ is the characteristic function of the random variable W and the random variables of the sequence $\{X_n, n=1,2,\ldots\}$ are equally distributed with the random variable X having characteristic function $\varphi_X(u)$ then (3.7.14) has the form

$$\varphi_{T,W}(u,\xi) = \varphi_W(\xi)\sum_{n=1}^{\infty} \varphi_X^n(u)pq^{n-1}$$

or equivalently the form

$$\varphi_{T,W}(u,\xi) = \varphi_W(\xi)\varphi_X(u)p\sum_{n=1}^{\infty}(q\varphi_X(u))^{n-1}. \tag{3.7.15}$$

From (3.7.2), (3.7.4) and (3.7.15) we get that

$$\varphi_{T,W}(u,\xi) = \frac{p\varphi_X(u)}{1-q\varphi_X(u)}\varphi_W(\xi)$$

or equivalently we get that

$$\varphi_{T,W}(u,\xi) = \varphi_T(u)\varphi_W(\xi). \tag{3.7.16}$$

From (3.7.16) we conclude that the random sum

$$T = X_1 + X_2 + \cdots + X_N$$

is independent of the random variable W.

If $\varphi_V(u)$ is the characteristic function of the random variable in (3.7.8) $V = T\,W$ or equivalently the random variable

$$V = (X_1 + X_2 + \cdots + X_N)W$$

then we have

$$\varphi_V(u) = E(e^{iuV})$$

or equivalently we have

$$\varphi_V(u) = E\big(E(e^{iuV}|W)\big). \tag{3.7.17}$$

Since $f_W(w)$ is the probability density function of the random variable W then (3.7.17) implies that

$$\varphi_V(u) = \int_0^{\infty} E(e^{iuV}|W=w)f_W(w)dw$$

or equivalently

$$\varphi_V(u) = \int_0^{\infty} E(e^{iuTW}|W=w)f_W(w)dw. \tag{3.7.18}$$

Hence (3.7.18) implies that

$$\varphi_V(u) = \int_0^\infty E(e^{iuwT}|W = w)f_W(w)dw. \tag{3.7.19}$$

Since the random variable

$$T = X_1 + X_2 + \cdots + X_N$$

is independent of the random variable W then (3.7.19) implies that

$$\varphi_V(u) = \int_0^\infty E(e^{iuwT})f_W(w)dw. \tag{3.7.20}$$

Since

$$\varphi_T(uw) = E(e^{iuwT})$$

and

$$\varphi_T(uw) = \frac{p\varphi_X(uw)}{1 - q\varphi_X(uw)}$$

then (3.7.20) implies that the characteristic function $\varphi_V(u)$ of the random variable $V = TW$ or equivalently the random variable

$$V = (X_1 + X_2 + \cdots + X_N)W$$

is

$$\varphi_V(u) = \int_0^\infty \frac{p\varphi_X(uw)}{1 - q\varphi_X(uw)} f_W(w)dw. \tag{3.7.21}$$

If the random variables of the sequence $\{X_n, n = 1, 2, \ldots\}$ follow the exponential distribution with characteristic function

$$\varphi_X(u) = \frac{\mu}{\mu - iu}$$

and the random variable W follows the uniform distribution with probability density function

$$f_W(w) = 1$$

then (3.7.21) implies that the characteristic function $\varphi_V(u)$ of the random variable

$$V = (X_1 + X_2 + \cdots + X_N)W$$

has the form

$$\varphi_V(u) = \int_0^1 \frac{p\mu}{p\mu - iuw} dw$$

or equivalently the form

$$\varphi_V(u) = \frac{p\mu}{iu} \log\left(\frac{p\mu}{p\mu - iu}\right).$$

The random variable

$$V = (X_1 + X_2 + \cdots + X_N)W,$$

denoting the cost of applying the risk frequency reduction operation up to the first retained risk occurrence, is very significant for selecting this operation. The random variable J, denoting the economic benefit of applying this operation, is also very significant for selecting the operation. The formulation of the random variable J is implemented by the following way.

Let $\{C_n, n = 1, 2, \ldots\}$ be a sequence of continuous, positive, independent, and identically distributed random variables. The random variables of the sequence are equally distributed with the random variable C having characteristic function $\varphi_C(u)$.

The random variable $C_n, n = 1, 2, \ldots$ denotes the size of damage due to the nth risk occurrence.

We consider the discrete random variable K denoting the number of deleted risk occurrences up to the first retained risk occurrence. The random variable K follows the geometric type I distribution with probability generating function

$$\varphi_K(z) = \frac{p}{1 - qz}.$$

The random variable K is independent of the sequence $\{C_n, n = 1, 2, \ldots\}$.

Hence the random variable J, denoting the economic benefit of applying the risk frequency reduction operation up to the first retained risk occurrence, has the form

$$J = C_1 + C_2 + \cdots + C_K$$

□

3.8 Capital of Treatment of Ongoing Risk Occurrences

Let $\{N(t), t \geq 0\}$ be a homogeneous Poisson process with $E(N(t)) = \lambda t$.

We suppose that the random variable $N(t)$ denotes the frequency of a risk in the time interval $[0,\ t]$ and $\{Y_n, n = 1, 2, \ldots\}$ is a sequence of continuous, positive, independent, and identically distributed random variables. The random variables of the sequence $\{Y_n, n = 1, 2, \ldots\}$ denote the durations of the occurrences of the risk in the time interval $[0,\ t]$ and these random variables are equally distributed with the random variable Y having distribution function $F_Y(y)$.

If the random variable $\Pi(t)$ denotes the number of risk occurrences in the time interval $[0,\ t]$ which are ongoing at the time point t then Theorem 2.9.2 implies that the random variable $\Pi(t)$ follows the Poisson distribution with probability generating function

$$P_{\Pi(t)}(z) = e^{\lambda pt(z-1)} \tag{3.8.1}$$

where

$$p = \int_0^t \frac{1 - F_Y(y)}{y} dy.$$

We consider the sequence of continuous, positive, independent, and identically distributed random variables $\{X_\pi, \pi = 1, 2, \ldots\}$.

The random variables of the sequence $\{X_\pi, \pi = 1, 2, \ldots\}$ are equally distributed with the random variable X having characteristic function

$$\varphi_X(u). \tag{3.8.2}$$

The random variable X_π denotes the capital employed for treating, at the time point t, the πth risk occurrence arising in the time interval $[0,\ t]$ and which is ongoing at the time point t.

The random variable $\Pi(t)$ is independent of the sequence $\{X_\pi, \pi = 1, 2, \ldots\}$.

The random sum

$$H = X_1 + X_2 + \cdots + X_{\Pi(t)}$$

denotes the capital employed for treating, at the time point t, the occurrences of the risk in the time interval $[0,\ t]$ and which are ongoing at the time point t.

The investigation of the random sum

$$H = X_1 + X_2 + \cdots + X_{\Pi(t)}$$

is implemented by the use of the corresponding characteristic function $\varphi_{\Pi(t)}(u)$.

From (2.5.6), (3.8.1) and (3.8.2) it follows that the characteristic function of the random sum

$$H = X_1 + X_2 + \cdots + X_{\Pi(t)}$$

is

$$\varphi_H(u) = e^{\lambda pt(\varphi_X(u)-1)}. \tag{3.8.3}$$

Special cases of the characteristic function (3.8.3) are the following.

We suppose that the random variable X follows the uniform distribution with characteristic function

$$\varphi_X(u) = \frac{e^{iu} - 1}{iu}. \tag{3.8.4}$$

From (3.8.3) and (3.8.4) it follows that the characteristic function of the random sum

$$H = X_1 + X_2 + \cdots + X_{\Pi(t)}$$

has the form

$$\varphi_H(u) = \exp\left[\lambda pt\left(\frac{e^{iu} - 1}{iu} - 1\right)\right].$$

We suppose that the random variable X follows the exponential distribution with characteristic function

$$\varphi_X(u) = \frac{\mu}{\mu - iu}. \tag{3.8.5}$$

From (3.8.3) and (3.8.5) it follows that the characteristic function of the random sum

$$H = X_1 + X_2 + \cdots + X_{\Pi(t)}$$

has the form

$$\varphi_H(u) = \exp\left[\lambda pt\left(\frac{\mu}{\mu - iu} - 1\right)\right].$$

From a theoretical point of view the characteristic function

$$\varphi_H(u) = e^{\lambda pt(\varphi_X(u)-1)}$$

of the random sum

$$H = X_1 + X_2 + \cdots + X_{\Pi(t)}$$

consists a very strong tool for investigating the class of infinitely divisible probability distributions. This class includes the class of stable probability distributions and the class of self-decomposable probability distributions. These classes have many important applications in various areas of probability theory. Moreover, these classes are particularly useful in formulating stochastic models for processes and systems arising in significant practical disciplines. From a practical point of view the characteristic function

$$\varphi_H(u) = e^{\lambda pt(\varphi_X(u)-1)}$$

consists a fundamental factor for the applications of the random sum

$$H = X_1 + X_2 + \cdots + X_{\Pi(t)}$$

in various practical disciplines. These applications are based on the theoretical properties of the probability distribution of the random sum

$$H = X_1 + X_2 + \cdots + X_{\Pi(t)}$$

The establishment of properties of the above random sum is implemented by the use of the corresponding characteristic function

$$\varphi_H(u) = e^{\lambda pt(\varphi_X(u)-1)}.$$

The contribution of the present section consists of the introduction of the concept of the random sum

$$H = X_1 + X_2 + \cdots + X_{\Pi(t)}$$

This random sum can be considered as a structural element for treatment of a risk which is ongoing at the given time point t.

The probabilistic information which is included in the corresponding characteristic function

$$\varphi_{\mathrm{H}}(u) = e^{\lambda pt(\varphi_X(u)-1)}$$

supports the development of a proactive risk management program facilitating the treatment of a given risk. Such a program incorporates the formulation of a random sum of the form

$$H = X_1 + X_2 + \cdots + X_{\Pi(t)}$$

for every risk threatening an organization in the time interval $[0, t]$.

The formulations of these random sums are possible if the risks threatening an organization satisfy the conditions of the present section. In this case the capital employed for treating, at the time point t of the risk occurrences arising in the time interval $[0, t]$ and which are ongoing at this time point t, is represented by a sum of random sums of the form

$$H = X_1 + X_2 + \cdots + X_{\Pi(t)}$$

If the risks threatening an organization are independent then the characteristic function of the capital of treating all the ongoing risk occurrences at the time point t is a product of characteristic functions of the form

$$\varphi_H(u) = e^{\lambda pt(\varphi_X(u)-1)}.$$

3.9 Binomial Random Sums in Modeling Risk Control Operations

The occurrences of different risks threatening an activity make very difficult the implementation of the purposes of the activity. An investment is an activity having as main purpose the creation of income of a given size. The difficulties of implementing that purpose are due to the occurrences of some risks threatening the investment. It is a common financial practice the use of the market value of an investment for financing other investments. The present section concentrates on the formulation of a random sum for the investigation of such financial practices. We consider a set including n similar and independent investments and the finite sequence of continuous, positive, independent, and identically distributed random variables $\{X_\kappa, \kappa = 1, 2, \ldots, n\}$.

The random variables of the above sequence are equally distributed with the random variable X having distribution function $F_X(x)$.

The random variable $X_\kappa, \kappa = 1, 2, \ldots, n$ denotes the income generated by the κth, $\kappa = 1, 2, \ldots, n$ investment in the time interval $[0, t]$.

We consider the positive real number c. If the income $X_\kappa, \kappa = 1, 2, \ldots, n$ generated by the κth, $\kappa = 1, 2, \ldots, n$ investment in the time interval $[0, t]$ is smaller than the positive real number c, or equivalently the event $(X_K < c), \kappa = 1, 2, \ldots, n$ occurs then the investment κth, $\kappa = 1, 2, \ldots, n$ is removed from the set of investments. The probability of the event $(X_K < c)$ is $p = P(X_K < c)$ or equivalently $p = F_X(c)$.

Let N be a discrete random variable denoting the number of investments which will be removed from the set of investments at the end of the time interval $[0,\ t]$.

Since the investments of the set are independent then the discrete random variable N follows the binomial distribution with probability function

$$P(N=\kappa)=\binom{n}{\kappa}p^{\kappa}q^{n-\kappa},\quad \kappa=0,1,2,\ldots,n$$

and probability generating function

$$P_N(z)=(pz+q)^n. \tag{3.9.1}$$

We consider the finite sequence of continuous, positive, independent, and identically distributed random variables $\{V_\kappa, \kappa=1,2,\ldots,n\}$.

The random variables of the above sequence are equally distributed with the random variable V having characteristic function

$$\varphi_V(u). \tag{3.9.2}$$

The random variable $V_\kappa, \kappa=1,2,\ldots,n$ denotes the market value of the κth, $\kappa=1,2,\ldots,n$ investment at the end of the time interval $[0,\ t]$.

The random variable N is independent of the sequence of continuous, positive, independent, and identically distributed random variables $\{V_\kappa, \kappa=1,2,\ldots,n\}$.

The binomial random sum

$$Y=V_1+V_2+\cdots+V_N$$

denotes the total market value of the investments which will be removed from the set of investments at the end of the time interval $[0,\ t]$.

Moreover, the binomial random sum

$$Y=V_1+V_2+\cdots+V_N$$

is a stochastic model with very useful applications in investment decision making. From Sect. 2.5, (3.9.1) and (3.9.2) it follows that the characteristic function $\varphi_Y(u)$ of the random sum

$$Y=V_1+V_2+\cdots+V_N$$

has the form

$$\varphi_Y(u)=(p\varphi_V(u)+q)^n.$$

The establishment of properties of the random sum

$$Y=V_1+V_2+\cdots+V_N,$$

with practical and theoretical interest, can be implemented by the use of the characteristic function

$$\varphi_Y(u) = (p\varphi_V(u) + q)^n.$$

Moreover, the establishment of sufficient conditions for embedding the characteristic function

$$\varphi_Y(u) = (p\varphi_V(u) + q)^n$$

in significant classes of characteristic functions is particularly useful for investigating the probabilistic behavior of the binomial random sum

$$Y = V_1 + V_2 + \cdots + V_N.$$

A very important element of formulating the binomial random sum

$$Y = V_1 + V_2 + \cdots + V_N$$

is the selection of the point t of the time interval $[0,\ t]$.

The selection of the time point t mainly depends on two factors. The first factor is the financial strategy of the firm having the set of the n similar and independent investments. The second factor is the set of risks threatening the set of the n similar and independent investments.

Since the random variable $V_\kappa, \kappa = 1, 2, \ldots, n$ denotes the market value of the κth, $\kappa = 1, 2, \ldots, n$ investment at the end of the time interval $[0,\ t]$ then the time point t is smaller than the duration of every investment of the set containing n similar and independent investments.

The mathematical structure of the binomial random sum

$$Y = V_1 + V_2 + \cdots + V_N$$

makes possible the financing of new investments by incorporating the total market value of a set of random and binomially distributed number of investments, with the income of an investment of that set in the time interval $[0,\ t]$ to be smaller than the positive real number c.

Since the occurrences of some risks can make an investment to generate income smaller than the positive real number c in the time interval $[0,\ t]$ then the binomial random sum

$$Y = V_1 + V_2 + \cdots + V_N$$

and the corresponding characteristic function

$$\varphi_Y(u) = (p\varphi_V(u) + q)^n$$

can be considered as analytical tools of risk management with interesting practical applications.

Other applications of the binomial random sum

$$Y = V_1 + V_2 + \cdots + V_N$$

in the area of risk financing operations can be established by providing the random variables of the finite sequence $\{V_\kappa, \kappa = 1, 2, \ldots, n\}$ with additional interpretations in various areas of the discipline of risk management.

3.10 Modeling Risk Financing Operations

The purpose of the present section is the formulation and investigation of a stochastic model for the description and analysis of an activity for financing a risk faced by a firm.

We consider a sequence of successive and equal time intervals. Let $\{X_n, n = 1, 2, \ldots\}$ be a sequence of independent and equally distributed random variables. The random variables of the sequence are equally distributed with the random variable X which follows the Bernoulli distribution with probability function

$$P(X = x) = \begin{cases} \theta, & x = 0 \\ \omega, & x = 1. \end{cases} \tag{3.10.1}$$

The random variable $\{X_n, n = 1, 2, \ldots\}$ denotes the frequency of the risk in the nth time interval. The organization saves the amount 1 in the beginning of every time interval in order to create a reserve for financing the impacts of risk occurrences. If the risk does not occur in the nth time interval then the firm spends the reserve which has been created until the beginning of the nth time interval. If the risk occurs in the nth time interval then the firm maintains the reserve which has been created until the beginning of this time interval. We consider the sequence of random variables $\{Y_n, n = 1, 2, \ldots\}$.

The random variable $Y_n, n = 1, 2, \ldots$ denotes the reserve until the beginning of the nth time interval. In the beginning of the first time interval the reserve is $Y_0 = 1$.

From the definition of the random variable $X_n, n = 1, 2, \ldots$, the definition of the random variable $Y_n, n = 1, 2, \ldots$ and the description of the process of creating the reserve for financing the damages due to the occurrences of the risk we get that

$$Y_n = 1 + X_n Y_{n-1}, \quad n = 1, 2, \ldots. \tag{3.10.2}$$

The investigation of the stochastic model (3.10.2) can be implemented by the use of the corresponding characteristic function $\varphi_{Y_n}(u)$.

The following result concentrates on the evaluation of the characteristic function $\varphi_{Y_n}(u)$ and the characteristic function $\lim_{n\to\infty} \varphi_{Y_n}(u)$.

Theorem 3.10.1 *Let $\{X_n, n = 1, 2, \ldots\}$ be a sequence of independent and identically distributed random variables. The random variables of the sequence are equally distributed with the random variable X following the Bernoulli distribution with probability function*

$$P(X = x) = \begin{cases} \theta, & x = 0 \\ \omega, & x = 1. \end{cases}$$

We consider the sequence of random variables $\{Y_n, n = 1, 2, \ldots\}$ with $Y_0 = 1$ and

$$Y_n = 1 + X_n Y_{n-1}, \quad n = 1, 2, \ldots.$$

If $\varphi_{Y_n}(u)$ is the characteristic function of the random variable Y_n then

$$\varphi_{Y_n}(u) = \theta e^{iu} \sum_{m=1}^{n} \omega^{m-1} e^{i(m-1)u} + \omega^n e^{i(n+1)u}$$

or equivalently

$$\varphi_{Y_n}(u) = \theta e^{iu} \frac{\omega^n e^{inu} - 1}{\omega e^{iu} - 1} + \omega^n e^{i(n+1)u}$$

and

$$\lim_{n \to \infty} \varphi_{Y_n}(u) = \frac{\theta e^{iu}}{1 - \omega e^{iu}}.$$

Proof The evaluation of the characteristic function $\varphi_{Y_n}(u)$ of the random variable Y_n will be implemented by mathematical induction.

We have $\varphi_{Y_n}(u) = E(e^{iuY_n})$ or equivalently

$$\varphi_{Y_n}(u) = E\big(E\big(e^{iuY_n} | X_n\big)\big). \tag{3.10.3}$$

From (3.10.3) it follows that

$$\varphi_{Y_n}(u) = \sum_{x=0}^{1} E\big(e^{iuY_n} | X_n = x\big) P(X_n = x)$$

or equivalently

$$\varphi_{Y_n}(u) = \sum_{x=0}^{1} E\left(e^{iu(1+X_nY_{n-1})}|X_n = x\right)P(X_n = x). \tag{3.10.4}$$

From (3.10.4) we get that

$$\varphi_{Y_n}(u) = \sum_{x=0}^{1} E(e^{iu}e^{iuX_nY_{n-1}}|X_n = x)P(X_n = x)$$

or equivalently we get that

$$\varphi_{Y_n}(u) = e^{iu}\sum_{x=0}^{1} E(e^{iuxY_{n-1}}|X_n = x)P(X_n = x). \tag{3.10.5}$$

Since the random variable Y_{n-1} is independent of the random variable X_n then (3.10.5) has the form

$$\varphi_{Y_n}(u) = e^{iu}\sum_{x=0}^{1} E(e^{iuxY_{n-1}})P(X_n = x) \tag{3.10.6}$$

From (3.10.1) and (3.10.6) it follows that

$$\varphi_{Y_n}(u) = e^{iu}(\theta + \omega E(e^{iuY_{n-1}})). \tag{3.10.7}$$

Since $\varphi_{Y_{n-1}}(u) = E(e^{iuY_{n-1}})$ is the characteristic function of the random variable Y_{n-1} then (3.10.7) has the form

$$\varphi_{Y_n}(u) = e^{iu}(\theta + \omega\varphi_{Y_{n-1}}(u)). \tag{3.10.8}$$

The relationship (3.10.8) will be used to prove by induction that the characteristic function of the random Y_n has the form

$$\varphi_{Y_n}(u) = \theta e^{iu}\sum_{m=1}^{n} \omega^{m-1}e^{i(m-1)u} + \omega^n e^{i(n+1)u}. \tag{3.10.9}$$

If $n = 1$ then (3.10.9) has the form

$$\varphi_{Y_1}(u) = e^{iu}(\theta + \omega e^{iu}). \tag{3.10.10}$$

For $n = 1$ then (3.10.2) has the form $Y_1 = 1 + X_1$, since $Y_0 = 1$.
The characteristic function of the random variable $Y_1 = 1 + X_1$ is

$$\varphi_{Y_1}(u) = E\left(e^{iu(1+X_1)}\right)$$

or equivalently

$$\varphi_{Y_1}(u) = e^{iu}E\left(e^{iuX_1}\right). \tag{3.10.11}$$

From (3.10.1) and (3.10.11) it follows that

$$\varphi_{Y_1}(u) = e^{iu}\left(\theta + \omega e^{iu}\right). \tag{3.10.12}$$

Moreover (3.10.10) and (3.10.12) imply that (3.10.9) is valid for $n = 1$. We suppose that (3.10.9) is valid for $n = \kappa$.

Hence

$$\varphi_{Y_\kappa}(u) = \theta e^{iu}\sum_{m=1}^{\kappa}\omega^{m-1}e^{i(m-1)u} + \omega^{\kappa}e^{i(\kappa+1)u}. \tag{3.10.13}$$

From (3.10.8) if $n = \kappa + 1$ we get that

$$\varphi_{Y_{\kappa+1}}(u) = e^{iu}\left(\theta + \omega\varphi_{Y_\kappa}(u)\right). \tag{3.10.14}$$

From (3.10.13) and (3.10.14) it follows that

$$\varphi_{Y_{\kappa+1}}(u) = e^{iu}\left\{\theta + \omega\left[\theta e^{iu}\sum_{m=1}^{\kappa}\omega^{m-1}e^{i(m-1)u} + \omega^{\kappa}e^{i(\kappa+1)u}\right]\right\}$$

or equivalently

$$\varphi_{Y_{\kappa+1}}(u) = \theta e^{iu}\sum_{m=1}^{\kappa+1}\omega^{m-1}e^{i(m-1)u} + \omega^{\kappa+1}e^{i(\kappa+2)u}. \tag{3.10.15}$$

From (3.10.15) it follows that (3.10.9) is valid for $n = \kappa + 1$. Hence the characteristic function of the random variable $Y_n, n = 1, 2, \ldots$ has the form

$$\varphi_{Y_n}(u) = \theta e^{iu}\sum_{m=1}^{n}\omega^{m-1}e^{i(m-1)u} + \theta^{n}e^{i(n+1)u}. \tag{3.10.16}$$

The limiting behaviour of the sequence $\{Y_n, n = 1, 2, \ldots\}$ is of particular theoretical and practical interest.

From (3.10.16) it follows

$$\varphi_{Y_n}(u) = \theta e^{iu} \frac{\omega^n e^{inu} - 1}{\omega e^{iu} - 1} + \omega^n e^{i(n+1)u}. \quad (3.10.17)$$

Since

$$e^{iu} = \cos u + i \sin u$$

and $|e^{iu}| = 1$ then from (3.10.17) it follows that

$$\lim_{n \to \infty} \varphi_{Y_n}(u) = \frac{\theta e^{iu}}{\omega e^{iu} - 1} \lim_{n \to \infty} \left(\omega^n e^{inu} - 1\right) + \lim_{n \to \infty} \omega^n e^{i(n+1)u}$$

or equivalently

$$\lim_{n \to \infty} \varphi_{Y_n}(u) = \frac{\theta e^{iu}}{1 - \omega e^{iu}}. \quad (3.10.18)$$

From (3.10.18) it follows that the sequence of random variables $\{Y_n, n = 1, 2, \ldots\}$ converges to the random variable Y following the geometric type II distribution with probability function

$$P(Y = y) = \theta \omega^{y-1}, \quad y = 1, 2, \ldots.$$

The theoretical and practical distribution of the present section consists of interpreting the role of the stochastic model

$$Y_n = 1 + X_n Y_{n-1}, \quad n = 1, 2, \ldots$$

in the area of theoretical and practical applications of risk financing operations. □

3.11 Time of First Retained Risk Occurrence and Total Cost of Deleting Risk Occurrences

Let N be a discrete random variable with values in the set $\mathbf{N} = \{1, 2, \ldots\}$ following the geometric type II distribution with probability generating function

$$P_N(z) = \frac{pz}{1 - qz}.$$

We suppose that $\{X_n, n = 1, 2, \ldots\}$ is a sequence of continuous, positive, independent, and identically distributed random variables. The random variables of the sequence are equally distributed with the random variable X having characteristic function $\varphi_X(u)$ and we set

$$T = X_1 + X_2 + \cdots + X_N.$$

We suppose that $\{C_n, n = 1, 2, \ldots\}$ is a sequence of continuous, positive, independent, and identically distributed random variables. The random variables of the sequence are equally distributed with the random variable C having characteristic function $\varphi_C(\xi)$ and we set

$$L = C_1 + C_2 + \cdots + C_N.$$

We consider the vector (T, L). The purpose of the present section is the establishment of properties and applications in the discipline of risk management of the vector (T, L).

The following result establishes sufficient and necessary conditions for the evaluation of the characteristic function $\varphi_{T,L}(u, \xi)$ of the vector (T, L).

Theorem 3.11.1 *Let N be a discrete random variable with values in the set $\mathbf{N} = \{1, 2, \ldots\}$ following the geometric type II distribution with probability generating function*

$$P_N(z) = \frac{pz}{1 - qz}.$$

We suppose that $\{X_n, n = 1, 2, \ldots\}$ is a sequence of continuous, positive, independent, and identically distributed random variables. The random variables of the sequence are equally distributed with the random variable X having characteristic function $\varphi_X(u)$ and we set

$$T = X_1 + X_2 + \cdots + X_N.$$

We suppose that $\{C_n, n = 1, 2, \ldots\}$ is a sequence of continuous, positive, independent, and identically distributed random variables. The random variables of the sequence are equally distributed with the random variable C having characteristic function $\varphi_C(\xi)$ and we set

$$L = C_1 + C_2 + \cdots + C_N.$$

If N, $\{X_n, n = 1, 2, \ldots\}$ and $\{C_n, n = 1, 2, \ldots\}$ are independent then the characteristic function of the vector (T, L) is

$$\varphi_{T,L}(u, \xi) = \frac{p\varphi_X(u)\varphi_C(\xi)}{1 - q\varphi_X(u)\varphi_C(\xi)}.$$

Proof The proof of Theorem 3.11.1 follows from the proof of Theorem 2.12.1.

An interpretation of the vector (T, L), where

$$T = X_1 + X_2 + \cdots + X_N$$

and

$$L = C_1 + C_2 + \cdots + C_N,$$

in the discipline of risk management is the following.

We consider the sequence $\{X_n, n = 1, 2, \ldots\}$ of continuous, positive, independent, and identically distributed random variables, and we suppose that the random variable X_n denotes the time between the nth and the $(n-1)$th occurrence of a risk. A risk frequency reduction operation is applied to the risk. This operation retains with probability p and deletes with probability $q = 1 - p$ a risk occurrence. The retention–deletion of a risk occurrence is independent of the retention–deletion of any other risk occurrence. Let N be a random variable denoting the number of the risk occurrences until the first retained risk occurrence. The random variable N follows the geometric type II distribution with probability generating function

$$P_N(z) = \frac{pz}{1 - qz}.$$

Hence the random variable

$$T = X_1 + X_2 + \cdots + X_N$$

denotes the time of the first retained risk occurrence. We consider the sequence $\{C_n, n = 1, 2, \ldots\}$ of continuous, positive, independent, and identically distributed random variables, and we suppose that the random variable C_n denotes the cost of applying the risk frequency reduction operation to the nth risk occurrence. Hence the random variable

$$L = C_1 + C_2 + \cdots + C_N$$

denotes the total cost of applying the risk frequency reduction operation until the first retained risk occurrence. The random variable

$$T = X_1 + X_2 + \cdots + X_N,$$

the random variable

$$L = C_1 + C_2 + \cdots + C_N,$$

and the vector (T, L) are strong analytical tools for investigating and applying risk frequency reduction operations.

The independence of N, $\{X_n, n = 1,2,\ldots\}$ and $\{C_n, n = 1,2,\ldots\}$ permits the evaluation of the characteristic function

$$\varphi_{T,L}(u,\xi) = \frac{p\varphi_X(u)\varphi_C(\xi)}{1 - q\varphi_X(u)\varphi_C(\xi)}$$

of the vector (T,L).

That characteristic function supports the practical applications of the above random sum in risk management. □

3.12 Occurrence Time and Total Severity of First Retained Risk Occurrence

Let N be a discrete random variable with values in the set $\mathbf{N} = \{1,2,\ldots\}$ and following the geometric type II distribution with probability generating function

$$P_N(z) = \frac{pz}{1-qz}.$$

We consider the sequence $\{X_n, n = 1,2,\ldots\}$ of continuous, positive, independent, and identically distributed random variables. The random variables of the sequence are equally distributed with the random variable X having characteristic function $\varphi_X(u)$ and we set

$$T = X_1 + X_2 + \cdots + X_N.$$

Let S be a discrete random variable with values in the set $\mathbf{N}_0 = \{0,1,2,\ldots\}$ and probability generating function $P_S(z)$.

We consider the sequence $\{C_s, s = 1,2,\ldots\}$ of continuous, positive, and independent random variables. The random variables of the sequence are equally distributed with the random variable C having characteristic function $\varphi_C(\xi)$ and we set

$$L = C_1 + C_2 + \cdots + C_S.$$

We consider the vector (T,L). The purpose of the present section is the establishment of properties and applications in the discipline of risk management of the above vector.

The following result establishes sufficient conditions for evaluating the characteristic function $\varphi_{T,L}(u,\xi)$ of the vector (T,L).

Theorem 3.12.1 *Let N be a discrete random variable with values in the set $\mathbf{N} = \{1,2,\ldots\}$ and following the geometric type II distribution with probability generating function*

$$P_N(z) = \frac{pz}{1 - qz}.$$

We consider the sequence $\{X_n, n = 1, 2, \ldots\}$ *of continuous, positive, independent, and identically distributed random variables. The random variables of the sequence are equally distributed with the random variable* X *having characteristic function* $\varphi_X(u)$ *and we set*

$$T = X_1 + X_2 + \cdots + X_N.$$

Let S *be a discrete random variable with values in the set* $\mathbf{N}_0 = \{0, 1, 2, \ldots\}$ *and probability generating function* $P_S(z)$.

We consider the sequence $\{C_s, s = 1, 2, \ldots\}$ *of continuous, positive, independent, and identically distributed random variables. The random variables of the sequence are equally distributed with the random variable* C *having characteristic function* $\varphi_C(\xi)$ *and we set*

$$L = C_1 + C_2 + \cdots + C_S.$$

If N, $\{X_n, n = 1, 2, \ldots,\}$ S *and* $\{C_s, s = 1, 2, \ldots\}$ *are independent then the characteristic function of the vector* (T, L) *is*

$$\varphi_{T,L}(u, \xi) = \frac{p\varphi_X(u)}{1 - q\varphi_X(u)} P_S(\varphi_C(\xi)).$$

Proof The proof of Theorem 3.12.1 follows from the proof of Theorem 2.13.1.

An interpretation of the vector (T, L), where

$$T = X_1 + X_2 + \cdots + X_N$$

and

$$L = C_1 + C_2 + \cdots + C_S$$

in the discipline of risk management is the following.

We consider the sequence $\{X_n, n = 1, 2, \ldots\}$ of continuous, positive, independent, and identically distributed random variables, and we suppose that the random variable X_n denotes the time between the nth and the $(n - 1)$th occurrence of a risk. A risk frequency reduction operation is applied to the risk. This operation retains with probability p and deletes with probability $q = 1 - p$ a risk occurrence. The retention–deletion of a risk occurrence is independent of retention–deletion of any other risk occurrence. Let N be a random variable denoting the number of risk occurrences until the first retained risk occurrence. The random variable N follows the geometric type II distribution with probability generating function

$$P_N(z) = \frac{pz}{1 - qz}.$$

Hence the random variable

$$T = X_1 + X_2 + \cdots + X_N$$

denotes the time of the first retained risk occurrence. We suppose that the random variable S denotes the number of different damages due to the first retained risk occurrence at the time point

$$T = X_1 + X_2 + \cdots + X_N.$$

We consider the sequence $\{C_s, s = 1, 2, \ldots\}$ of continuous, positive, independent, and identically distributed random variables, and we suppose that the random variable C_S denotes the size of the sth damage due to the first retained risk occurrence. Hence the random variable

$$L = C_1 + C_2 + \cdots + C_S$$

denotes the total size of the damage due the first retained risk occurrence at the time point

$$T = X_1 + X_2 + \cdots + X_N.$$

The random variable

$$T = X_1 + X_2 + \cdots + X_N,$$

the random variable

$$L = C_1 + C_2 + \cdots + C_S$$

and the vector (T, L) are strong analytical tools for investigating and applying risk frequency reduction operations.

The independence of N, $\{X_n, n = 1, 2, \ldots,\}$ S and $\{C_s, s = 1, 2, \ldots\}$ permits the evaluation of the characteristic function

$$\varphi_{T,L}(u, \xi) = \frac{p\varphi_X(u)}{1 - q\varphi_X(u)} P_S(\varphi_C(\xi))$$

of the vector (T, L). That characteristic function supports the practical applications of the above random sum in risk management.

The probability distribution of the random vector (T, L) supports the investigation and selection of combinations of risk control and risk financing operations. Since the independence of N, $\{X_n, n = 1, 2, \ldots\}$, S and $\{C_s, s = 1, 2, \ldots\}$ implies the independence of N, $\{X_n, n = 1, 2, \ldots\}$ and the independence of S, $\{C_s, s = 1, 2, \ldots\}$, then the characteristic function of the random variable

$$T = X_1 + X_2 + \cdots + X_N$$

is

$$\varphi_T(u) = \frac{p\varphi_X(u)}{1 - q\varphi_X(u)}$$

and the characteristic function of the random variable

$$L = C_1 + C_2 + \cdots + C_S$$

is

$$\varphi_L(\xi) = P_S(\varphi_C(\xi)).$$

Hence the characteristic function

$$\varphi_{T,L}(u, \xi) = \frac{p\varphi_X(u)}{1 - q\varphi_X(u)} P_S(\varphi_C(\xi)),$$

the characteristic function

$$\varphi_T(u) = \frac{p\varphi_X(u)}{1 - q\varphi_X(u)}$$

and the characteristic function

$$\varphi_L(\xi) = P_S(\varphi_C(\xi))$$

satisfy the relationship

$$\varphi_{T,L}(u, \xi) = \varphi_T(u)\varphi_L(\xi).$$

Consequently, the random variables

$$T = X_1 + X_2 + \cdots + X_N$$

and

$$L = C_1 + C_2 + \cdots + C_S$$

are independent. □

3.13 Free of Risk Occurrences Time Interval and Total Benefit of Applying a Risk Frequency Reduction Operation

Let N be a discrete random random variable with values in the set $\mathbf{N} = \{1, 2, \ldots\}$ and following the geometric type II distribution with probability generating function

$$P_N(z) = \frac{pz}{1 - qz}.$$

We suppose that $\{X_n, n = 1, 2, \ldots\}$ is a sequence of continuous, positive, independent, and identically distributed random variables. The random variables of the sequence are equally distributed with the random variable X having characteristic function $\varphi_X(u)$ and we set

$$T = X_1 + X_2 + \cdots + X_N.$$

We consider the discrete random variable K with values in the set $\mathbf{N}_0 = \{0, 1, 2, \ldots\}$ and following the geometric type I distribution with probability generating function

$$P_K(z) = \frac{p}{1 - qz}$$

and we set

$$Y = X_1 + X_2 + \cdots + X_K.$$

We suppose that $\{C_\kappa, \kappa = 1, 2, \ldots\}$ is a sequence of continuous, positive, independent and identically distributed random variables. The random variables of the sequence are equally distributed with the random variable C having characteristic function $\varphi_C(\xi)$ and we set

$$V = C_1 + C_2 + \cdots + C_K.$$

We consider the vector (Y, V). The purpose of the present section is the establishment of properties and applications in risk management of the vector (Y, V).

The following result establishes sufficient conditions for evaluating the characteristic function $\varphi_{Y,V}(u, \xi)$ of the vector (Y, V).

Theorem 3.13.1 *Let N be a discrete random variable with values in the set $\mathbf{N} = \{1, 2, \ldots\}$ and following the geometric type II distribution with probability generating function*

$$P_N(z) = \frac{pz}{1 - qz}.$$

We suppose that $\{X_n, n = 1, 2, \ldots\}$ *is a sequence of continuous, positive, independent, and identically distributed random variables. The random variables of the sequence are equally distributed with the random variable* X *having characteristic function* $\varphi_X(u)$ *and we set*

$$T = X_1 + X_2 + \cdots + X_N.$$

We consider the discrete random variable K *with values in the set* $\mathbf{N}_0 = \{0, 1, 2, \ldots\}$ *and following the geometric type I distribution with probability generating function*

$$P_K(z) = \frac{p}{1 - qz}$$

and we set

$$Y = X_1 + X_2 + \cdots + X_K.$$

We consider the sequence $\{C_\kappa, \kappa = 1, 2, \ldots\}$ *of continuous, positive, and independent random variables. The random variables of the sequence are equally distributed with the random variable* C *having characteristic function* $\varphi_C(\xi)$ *and we set*

$$V = C_1 + C_2 + \cdots + C_K.$$

If K, $\{X_n, n = 1, 2, \ldots\}$ *and* $\{C_\kappa, \kappa = 1, 2, \ldots\}$ *are independent then the characteristic function of the vector* (Y, V) *is*

$$\varphi_{Y,V}(u, \xi) = \frac{p}{1 - q\varphi_X(u)\varphi_C(\xi)}.$$

Proof The proof of Theorem 3.13.1 follows from the proof of Theorem 2.12.1.

An interpretation of the vector (Y, V), where

$$Y = X_1 + X_2 + \cdots + X_K$$

and

$$V = C_1 + C_2 + \cdots + C_K,$$

in the discipline of risk management is the following.

We consider the sequence $\{X_n, n = 1, 2, \ldots\}$ of continuous, positive, independent, and identically distributed random variables and we suppose that the random variable X_n denotes the time between the nth and the $(n-1)$th risk occurrence. A risk frequency reduction operation is applied to the risk. That operation retains with probability p and deletes with probability $q = 1 - p$ a risk occurrence. The retention–deletion of a risk occurrence is independent of the retention–deletion of any other risk occurrence. Let N be a random variable denoting the number of risk occurrences until the first retained risk occurrence. The random variable N follows the geometric type II distribution with probability generating function

$$P_N(z) = \frac{pz}{1-qz}.$$

Hence the random variable

$$T = X_1 + X_2 + \cdots + X_N$$

denotes the time of the first retained risk occurrence.

Let K be the random variable denoting the number of deleted risk occurrences until the first retained risk occurrence. The random variable K follows the geometric type I distribution with probability generating function

$$P_K(z) = \frac{p}{1-qz}.$$

Hence the random variable

$$Y = X_1 + X_2 + \cdots + X_K$$

denotes the time required for the number of deleted risk occurrences until the first risk occurrence. That means that the time interval

$$[0,\ X_1 + X_2 + \cdots + X_K\,]$$

does not contain retained risk occurrences. We consider the sequence $\{C_\kappa, \kappa = 1, 2, \ldots\}$ of continuous, positive, independent, and identically distributed random variables, and the random variable C_κ denotes the size of damage due to the κ the risk occurrence. Since the time interval

$$[0,\ X_1 + X_2 + \cdots + X_K]$$

does not contain retained risk occurrences then the random variable

$$V = C_1 + C_2 + \cdots + C_K$$

denotes the benefit due to the application of the risk frequency reduction operation in the time interval

$$[0\,,\ X_1 + X_2 + \cdots + X_K\,]$$

The random variable

$$Y = X_1 + X_2 + \cdots + X_K,$$

the random variable

$$V = C_1 + C_2 + \cdots + C_K$$

and the vector (Y, V) are strong analytical tools for investigating and applying risk frequency reduction operations.

The independence of K, $\{X_n, n = 1, 2, \ldots\}$ and $\{C_\kappa, \kappa = 1, 2, \ldots\}$ permits the evaluation of the characteristic function

$$\varphi_{Y,V}(u, \xi) = \frac{p}{1 - q\varphi_X(u)\varphi_C(\xi)}$$

of the vector (Y, V).

That characteristic function supports the practical applications of the above vector in risk management. □

3.14 Free of Risk Occurrences Time Interval and Loan Portfolio

Let N be a discrete random variable with values in the set $\mathbf{N} = \{1, 2, \ldots\}$ and following the geometric type II distribution with probability generating function

$$P_N(z) = \frac{pz}{1 - qz}.$$

We suppose that $\{X_n, n = 1, 2, \ldots\}$ is a sequence of continuous, positive, independent, and identically distributed random variables. The random variables of the sequence are equally distributed with the random variable X having characteristic function $\varphi_X(u)$ and we set

$$T = X_1 + X_2 + \cdots + X_N.$$

We consider the discrete random variable K with values in the set $\mathbf{N}_0 = \{0, 1, 2, \ldots\}$ and following the geometric type I distribution with probability generating function

$$P_K(z) = \frac{p}{1-qz}$$

and set

$$Y = X_1 + X_2 + \cdots + X_K.$$

We suppose that S is a discrete random variable with values in the set $\mathbf{N}_0 = \{0,1,2,\ldots\}$ having probability generating function $P_S(z)$ and we consider the sequence $\{C_s, s = 1,2,\ldots\}$ of continuous, positive, independent, and identically distributed random variables. The random variables of the sequence are equally distributed with the random variable C having characteristic function $\varphi_C(\xi)$ and we set

$$V = C_1 + C_2 + \cdots + C_S.$$

We consider the vector (Y, V). The purpose of the present section is the establishment of properties and applications in risk management of the vector (Y, V).

The following result establishes sufficient conditions for evaluating the characteristic function $\varphi_{Y,V}(u,\xi)$ of the vector (Y, V).

Theorem 3.14.1 *Let N be a discrete random variable with values in the set $\mathbf{N} = \{1,2,\ldots\}$ and following the geometric type II distribution with probability generating function*

$$P_N(z) = \frac{pz}{1-qz}.$$

We suppose that $\{X_n, n = 1,2,\ldots\}$ is a sequence of continuous, positive, independent, and identically distributed random variables. The random variables of the sequence are equally distributed with the random variable X having characteristic function $\varphi_X(u)$ and we set

$$T = X_1 + X_2 + \cdots + X_N.$$

We consider the discrete random variable K with values in the set $\mathbf{N}_0 = \{0,1,2,\ldots\}$ and following the geometric type I distribution with probability generating function

$$P_K(z) = \frac{p}{1-qz}$$

and we set

$$Y = X_1 + X_2 + \cdots + X_K.$$

We consider the discrete random variable S with value in the set $\mathbf{N}_0 = \{0, 1, 2, \ldots\}$ *and probability generating function* $P_S(z)$ *and let* $\{C_s, s = 1, 2, \ldots\}$ *be a sequence of continuous, positive, independent, and identically distributed random variables. The random variables of the sequence are equally distributed with the random variable C having characteristic function* $\varphi_C(\xi)$ *and we set*

$$V = C_1 + C_2 + \cdots + C_S.$$

If K, $\{X_n, n = 1, 2, \ldots\}$, S *and* $\{C_s, s = 1, 2, \ldots\}$ *are independent then the characteristic function of the vector* (Y, V) *is*

$$\varphi_{Y,V}(u, \xi) = \frac{p}{1 - q\varphi_X(u)} P_S(\varphi_C(\xi)).$$

Proof The proof of Theorem 3.14.1 follows from the proof of the Theorem 2.13.1.

An interpretation of the vector (Y, V), where

$$Y = X_1 + X_2 + \cdots + X_K$$

and

$$V = C_1 + C_2 + \cdots + C_S$$

in risk management is the following.

A bank faces a risk at the time point 0. We consider the sequence $\{X_n, n = 1, 2, \ldots\}$ of continuous, positive, independent, and identically distributed random variables, and we suppose that the random variable X_n denotes the time between the nth and the $(n-1)$th risk occurrence. A risk frequency reduction operation is applied to the risk. That operation retains with probability p and deletes with probability $q = 1 - p$ a risk occurrence. The retention–deletion of a risk occurrence is independent of the retention–deletion of any other risk occurrence. Let N be a random variable denoting the number of risk occurrences until the first retained risk occurrence. The random variable N follows the geometric type II distribution with probability generating function

$$P_N(z) = \frac{pz}{1 - qz}.$$

Hence the random variable

$$T = X_1 + X_2 + \cdots + X_N$$

denotes the time of the first retained risk occurrence. Let K be a random variable denoting the number of deleted risk occurrences until the first retained risk occurrence. The random variable K follows the geometric type I distribution with probability generating function

$$P_K(z) = \frac{p}{1 - qz}.$$

Hence the random variable

$$Y = X_1 + X_2 + \cdots + X_K$$

denotes the time required for the deleted risk occurrences until the first retained risk occurrence. Consequently the random interval

$$[0, X_1 + X_2 + \cdots + X_K]$$

does not contain retained risk occurrences. We consider the discrete random variable S and we suppose that the above random variable denotes the number of loans that the bank provides in the time interval

$$[0, X_1 + X_2 + \cdots + X_K].$$

We suppose that the random variable C_s denotes the size of the sth loan that the bank provides in the time interval

$$[0, X_1 + X_2 + \cdots + X_K].$$

Hence the random variable

$$V = C_1 + C_2 + \cdots + C_S$$

denotes the size of the portfolio of loans that the bank creates in the time interval

$$[0, X_1 + X_2 + \cdots + X_K].$$

The random variable

$$Y = X_1 + X_2 + \cdots + X_K,$$

the random variable

$$V = C_1 + C_2 + \cdots + C_S,$$

and the vector (Y, V) constitute strong analytical tools for formulating and investigating portfolio of loans in a time interval which is free of risk occurrences and which has a random length.

The independence of K, $\{X_n, n = 1, 2, \ldots\}$ S and $\{C_s, s = 1, 2, \ldots\}$ permits the evaluation of the characteristic function

$$\varphi_{Y,V}(u,\xi) = \frac{p}{1 - q\varphi_X(u)} P_S(\varphi_C(\xi))$$

of the random vector (Y, V).

That characteristic function supports the practical and theoretical applications of the above vector in risk management.

Since the independence of K, $\{X_n, n = 1, 2, \ldots\}$, S and $\{C_s, s = 1, 2, \ldots\}$ implies the independence of K, $\{X_n, n = 1, 2, \ldots\}$ and the independence of S, $\{C_s, s = 1, 2, \ldots\}$, then the characteristic function of the random variable

$$Y = X_1 + X_2 + \cdots + X_K$$

is

$$\varphi_Y(u) = \frac{p}{1 - q\varphi_X(u)}$$

and the characteristic function of the random variable

$$V = C_1 + C_2 + \cdots + C_S$$

is

$$\varphi_V(\xi) = P_S(\varphi_C(\xi)).$$

Hence the characteristic function

$$\varphi_{Y,V}(u,\xi) = \frac{p}{1 - q\varphi_X(u)} P_S(\varphi_C(\xi)),$$

the characteristic function

$$\varphi_Y(u) = \frac{p}{1 - q\varphi_X(u)},$$

and the characteristic function

$$\varphi_V(\xi) = P_S(\varphi_C(\xi))$$

satisfy the relationship

$$\varphi_{Y,V}(u,\xi) = \varphi_Y(u)\varphi_V(\xi).$$

Consequently, the random variables

$$Y = X_1 + X_2 + \cdots + X_K$$

and

$$V = C_1 + C_2 + \cdots + C_S$$

are independent □

3.15 Cost of Deleting Occurrences of a Risk with Constant Probability and Total Severity of First Retained Risk Occurrence

Let N be a discrete random variable with values in the set $\mathbf{N} = \{1, 2, \ldots\}$ and following the geometric type II distribution with probability function

$$P(N = n) = pq^{n-1}, \quad n = 1, 2, \ldots$$

and probability generating function

$$P_N(z) = \frac{pz}{1 - qz}.$$

We suppose that $\{X_n, n = 1, 2, \ldots\}$ is a sequence of continuous, positive, independent, and identically distributed random variables. The random variables of the sequence are equally distributed with the random variable X having characteristic function $\varphi_X(u)$ and we set

$$T = X_1 + X_2 + \cdots + X_N.$$

Let W be a continuous and positive random variable with probability density function $f_W(w)$, $w > 0$ and we set

$$V = (X_1 + X_2 + \cdots + X_N)W$$

or equivalently $V = TW$.

Let S be a discrete random variable with values in the set $\mathbf{N}_0 = \{0, 1, 2, \ldots\}$ and probability generating function $P_S(z)$.

We suppose that $\{C_s, s = 1, 2, \ldots\}$ is a sequence of continuous, positive, independent, and identically distributed random variables. We suppose that the random variables of the sequence are equally distributed with the random variable C having characteristic function $\varphi_C(\xi)$ and we set

$$L = C_1 + C_2 + \cdots + C_S.$$

We consider the vector (V, L). The purpose of the present section is the establishment of properties and applications in risk management of the vector (V, L).

The following result establishes sufficient conditions for evaluating the characteristic function $\varphi_{V,L}(u, \xi)$ of the vector (V, L).

Theorem 3.15.1 *Let N be a discrete random variable with values in the set $\mathbf{N} = \{1, 2, \ldots\}$ and following the geometric type II distribution with probability function*

$$P(N = n) = pq^{n-1}, \quad n = 1, 2, \ldots \tag{3.15.1}$$

and probability generating function

$$P_N(z) = \frac{pz}{1 - qz}. \tag{3.15.2}$$

We suppose that $\{X_n, n = 1, 2, \ldots\}$ is a sequence of continuous, positive, independent and identically distributed random variables. The random variables of the sequence are equally distributed with the random variable X having characteristic function

$$\varphi_X(u) \tag{3.15.3}$$

and we set

$$T = X_1 + X_2 + \cdots + X_N.$$

Let W be a continuous and positive random variable with probability function

$$f_W(w), \quad w > 0 \tag{3.15.4}$$

and we set

$$V = (X_1 + X_2 + \cdots + X_N)W$$

or equivalently $V = TW$. Let S be a discrete random variable with values in the set $\mathbf{N}_0 = \{0, 1, 2, \ldots\}$ and probability generating function

$$P_S(z). \tag{3.15.5}$$

We suppose that $\{C_s, s = 1, 2, \ldots\}$ is a sequence of continuous, positive, independent, and identically distributed random variables. The random variables of the sequence are equally distributed with the random variable C having characteristic function

$$\varphi_C(\xi) \tag{3.15.6}$$

and we set

$$L = C_1 + C_2 + \cdots + C_S.$$

We consider the vector (V, L).

If N, $\{X_n, n = 1, 2, \ldots\}$ W, S and $\{C_s, s = 1, 2, \ldots\}$ are independent then the characteristic function of the vector (V, L) is

$$\varphi_{V,L}(u, \xi) = \int_0^\infty \frac{p\varphi_X(uw)}{1 - q\varphi_X(uw)} f_W(w)dw \cdot P_S(\varphi_C(\xi)).$$

Proof The independence of N $\{X_n, n = 1, 2, \ldots\}$, W, S and $\{C_s, s = 1, 2, \ldots\}$ implies the independence of N, $\{X_n, n = 1, 2, \ldots\}$ and the independence of S and $\{C_s, s = 1, 2, \ldots\}$.

Hence (2.5.6), (3.15.1), (3.15.2) and (3.15.3) imply that the characteristic function of the random sum

$$T = X_1 + X_2 + \cdots + X_N$$

is

$$\varphi_T(u) = \frac{p\varphi_X(u)}{1 - q\varphi_X(u)} \tag{3.15.7}$$

and (2.5.6), (3.15.5), (3.15.6) imply that the characteristic function of the random sum

$$L = C_1 + C_2 + \cdots + C_S$$

is

$$\varphi_L(\xi) = P_S(\varphi_C(\xi)). \tag{3.15.8}$$

From (3.15.4) it follows that the characteristic function of the random variable W is

$$\varphi_W(\vartheta) = \int_0^\infty e^{i\vartheta w} f_W(w)dw. \tag{3.15.9}$$

We consider the random variables

$$T = X_1 + X_2 + \cdots + X_N,$$

$$L = C_1 + C_2 + \cdots + C_S$$

and W and we shall prove that these random variables are independent. Let $\varphi_{T,L,W}(u, \xi, \vartheta)$ be the characteristic function of the vector (T, L, W).

The proof of independence of the random variables

$$T = X_1 + X_2 + \cdots + X_N,$$

$$L = C_1 + C_2 + \cdots + C_S$$

and W requires the proof of the relationship

$$\varphi_{T,L,W}(u, \xi, \vartheta) = \varphi_T(u)\varphi_L(\xi)\varphi_W(\vartheta).$$

We have

$$\varphi_{T,L,W}(u, \xi, \vartheta) = E\left(e^{iuT+i\xi L+i\vartheta W}\right)$$

or equivalently

$$\varphi_{T,L,W}(u, \xi, \vartheta) = E\left(e^{iu(X_1+X_2+\cdots+X_N)+i\xi(C_1+C_2+\cdots+C_s)+i\vartheta W}\right). \tag{3.15.10}$$

From (3.15.10) we get that

$$\varphi_{T,L,W}(u, \xi, \vartheta) = \sum_{n=1}^{\infty}\sum_{s=o}^{\infty} E\left(e^{iu(X_1+\cdots+X_N)+i\xi(C_1+\cdots+C_s)+i\vartheta W}|N = n, S = s\right)P(N = n, S = s)$$

or equivalently we get that

$$\varphi_{T,L,W}(u, \xi, \vartheta) = \sum_{n=1}^{\infty}\sum_{s=o}^{\infty} E\left(e^{iu(X_1+\cdots+X_n)+i\xi(C_1+\cdots+C_s)+i\vartheta W}|N = n, S = s\right)P(N = n, S = s). \tag{3.15.11}$$

From (3.15.11) it follows that

$$\varphi_{T,L,W}(u,\xi,\vartheta)=\sum_{n=1}^{\infty}\sum_{s=o}^{\infty}E(e^{iuX_1}\ldots e^{iuX_n}\cdot e^{iu\xi C_1}\ldots e^{iu\xi C_s+i\vartheta W}|N=n\,,\,S=s)P(N=n\,,\,S=s)$$

(3.15.12)

The independence of N, $\{X_n, n=1,2,\ldots\}$, W, S and $\{C_s, s=1,2,\ldots\}$ implies the independence of the random variables N, X_1, ..., X_n, W, S, C_1, ..., C_s and the independence of the random variables N, S.

Hence (3.15.12) has the form

$$\varphi_{T,L,W}(u,\xi,\vartheta)=\sum_{n=1}^{\infty}\sum_{s=o}^{\infty}E(e^{iuX_1}\ldots e^{iuX_n}\cdot e^{iu\xi C_1}\ldots e^{iu\xi C_s}e^{i\vartheta W})P(N-n)P(S-s).$$

(3.15.13)

The independence of the random variables N, X_1, ..., X_n, W, S, C_1, ..., C_s implies the independence of the random variables X_1, ..., X_n, W, C_1, ..., C_s.

Hence the random variables $e^{iuX_1},\ldots,e^{iuX_n}$, $e^{i\vartheta W}$, $e^{i\xi C_1},\ldots,e^{i\xi C_s}$ are independent and (3.15.13) has the form

$$\varphi_{T,L,W}(u,\xi,\vartheta)=E(e^{i\vartheta W})\sum_{n=1}^{\infty}E(e^{iuX_1})\ldots E(e^{iuX_n})P(N=n)\sum_{s=o}^{\infty}E(e^{i\xi C_1})\ldots E(e^{i\xi C_s})P(S=s).$$

(3.15.14)

Since the random variables of the sequence $\{X_n, n=1,2,\ldots\}$ are equally distributed with the random variable X having characteristic function $\varphi_X(u)$, the random variables of the sequence $\{C_s, s=1,2,\ldots\}$ are equally distributed with the random variable C having characteristic function $\varphi_C(\xi)$, and the random variable W has characteristic function $\varphi_W(\theta)$ then (3.15.14) has the form

$$\varphi_{T,L,W}(u,\xi,\vartheta)=\varphi_W(\vartheta)\sum_{n=1}^{\infty}\varphi_X^n(u)P(N=n)\sum_{s=o}^{\infty}\varphi_C^s(\xi)P(S=s)$$

or equivalently the form

$$\varphi_{T,L,W}(u,\xi,\vartheta)=\frac{p\varphi_X(u)}{1-q\varphi_X(u)}P_S(\varphi_C(\xi))\varphi_W(\vartheta). \tag{3.15.15}$$

From (3.15.7), (3.15.8), (3.15.9) and (3.15.15) it follows that

$$\varphi_{T,L,W}(u,\xi,\vartheta)=\varphi_T(u)\varphi_L(\xi)\varphi_W(\vartheta).$$

Hence the random variables

$$T = X_1 + X_2 + \cdots + X_N,$$

$$L = C_1 + C_2 + \cdots + C_S$$

and W are independent. The evaluation of the characteristic function $\varphi_{V,L}(u,\xi)$ of the vector (V, L), where

$$V = (X_1 + X_2 + \cdots + X_N)W$$

and

$$L = C_1 + C_2 + \cdots + C_S,$$

is implemented in the following way.

We have

$$\varphi_{V,L}(u,\xi) = E\left(e^{iuV+i\xi L}\right)$$

or equivalently

$$\varphi_{V,L}(u,\xi) = E\left(e^{iu(X+_1X_2+\cdots+X_N)W+i\xi(C_1+C_2+\cdots+C_S)}\right). \tag{3.15.16}$$

From (3.15.16) we get that

$$\varphi_{V,L}(u,\xi) = E\left(E\left(e^{iu(X_1+X_2+\cdots+X_N)W+i\xi(C_1+C_2+\cdots+C_S)}\right)|W\right)$$

Hence

$$\varphi_{V,L}(u,\xi) = \int_0^\infty E\left(e^{iu(X_1+X_2+\cdots+X_N)W+i\xi(C_1+C_2+\cdots+C_S)}|W=w\right)f_W(w)dw. \tag{3.15.17}$$

From (3.15.17) it follows that

$$\varphi_{V,L}(u,\xi) = \int_0^\infty E\left(e^{iu(X+_1X_2+\cdots+X_N)w+i\xi(C_1+C_2+\cdots+C_S)}|W=w\right)f_W(w)dw. \tag{3.15.18}$$

Since the random variables

$$T = X_1 + X_2 + \cdots + X_N,$$

$$L = C_1 + C_2 + \cdots + C_S$$

and W are independent then the random variables

$$wT = (X_1 + X_2 + \cdots + X_N)w,$$

$$L = C_1 + C_2 + \cdots + C_S$$

and W are independent. Hence (3.15.18) has the form

$$\varphi_{V,L}(u,\xi) = \int_0^\infty E\left(e^{iu(X_1+X_2+\cdots+X_N)w+i\xi(C_1+C_2+\cdots+C_S)}\right)f_W(w)dw. \tag{3.15.19}$$

The independence of the random variables wT, L and W implies the independence of the random variables

$$T = X_1 + X_2 + \cdots + X_N,$$

$$L = C_1 + C_2 + \cdots + C_S.$$

Hence (3.15.19) has the form

$$\varphi_{V,L}(u,\xi) = \int_0^\infty E\left(e^{iu(X+_1X_2+\cdots+X_N)w}\right)E\left(e^{i\xi(C+_1C_2+\cdots+C_S)}\right)f_W(w)dw. \tag{3.15.20}$$

From (3.15.20) it follows that

$$\varphi_{V,L}(u,\xi) = E\left(e^{i\xi(C_1+C_2+\cdots+C_S)}\right)\int_0^\infty E\left(e^{iu(X+_1X_2+\cdots+X_N)w}\right)f_W(w)dw. \tag{3.15.21}$$

Since (3.15.8) implies that the random variable

$$L = C_1 + C_2 + \cdots + C_S$$

has characteristic function

$$\varphi_L(\xi) = P_S(\varphi_C(\xi)),$$

then (3.15.21) has the form

$$\varphi_{V,L}(u,\xi) = P_S(\varphi_C(\xi)) \int_0^\infty E\left(e^{iu(X_1+X_2+\cdots+X_N)w}\right) f_W(w)dw. \tag{3.15.22}$$

Since (3.15.7) implies that the random variable

$$T = X_1 + X_2 + \cdots + X_N$$

has characteristic function

$$\varphi_T(u) = \frac{p\varphi_X(u)}{1 - q\varphi_X(u)}$$

then the random variable

$$wT = (X_1 + X_2 + \cdots + X_N)w$$

has characteristic function

$$\varphi_T(uw) = \frac{p\varphi_X(uw)}{1 - q\varphi_X(uw)}. \tag{3.15.23}$$

From (3.15.22) and (3.15.23) it follows that the characteristic function of the vector (V, L) is

$$\varphi_{V,L}(u,\xi) = \int_0^\infty \frac{p\varphi_X(uw)}{1 - q\varphi_X(uw)} f_W(w)dw \cdot P_S(\varphi_C(\xi)). \tag{3.15.24}$$

Since (3.7.21) implies that

$$\varphi_V(u) = \int_0^\infty \frac{p\varphi_X(uw)}{1 - q\varphi_X(uw)} f_W(w)dw$$

is the characteristic function of the random variable

$$V = (X_1 + X_2 + \cdots + X_N)W$$

then (3.15.24) implies that the random variable

$$V = (X_1 + X_2 + \cdots + X_N)W$$

is independent of the random variable

$$L = C_1 + C_2 + \cdots + C_s.$$

An interpretation of the vector (V, L), where

$$V = (X_1 + X_2 + \cdots + X_N)W$$

and

$$L = C_1 + C_2 + \cdots + C_S$$

in risk management is the following.

We consider an organization facing a risk at the time point 0. We suppose that $\{X_n, n = 1, 2, \ldots\}$ is a sequence of continuous, positive, independent, and identically distributed random variables. The random variables of the sequence are equally distributed with the random variable X having characteristic function $\varphi_X(u)$.

The random variable $X_n, n = 1, 2, \ldots$ denotes the time between the nth and the $(n-1)$th occurrence of a risk. A risk frequency reduction operation is applied to the risk. According to that operation a risk occurrence is retained with probability p and deleted with probability $q = 1 - p$.

The retention–deletion of a risk occurrence is independent of the retention–deletion of any other risk occurrence. Let N be a random variable denoting the number of risk occurrences until the first retained risk occurrence. The random variable N follows the geometric type II distribution with probability function

$$P(N = n) = pq^{n-1}, \quad n = 1, 2, \ldots$$

and probability generating function

$$P_N(z) = \frac{pz}{1 - qz}.$$

The sequence of continuous, positive, independent, and identically distributed random variables $\{X_n, n = 1, 2, \ldots\}$ is independent of the random variable N.

The random sum

$$T = X_1 + X_2 + \cdots + X_N$$

denotes the time of the first retained risk occurrence.

Let W be a continuous and positive random variable with probability density function $f_W(w)$.

We suppose that the random variable W denotes the cost per unit of time for applying the risk frequency reduction operation. Hence the random variable

$$V = (X_1 + X_2 + \cdots + X_N)W$$

denotes the cost of applying the risk frequency reduction operation until the first retained risk occurrence. We suppose that the random variable S denotes the number of different damages due to the first retained risk occurrence at the time point

$$T = X_1 + X_2 + \cdots + X_N.$$

The random variable S has probability generating function $P_S(z)$. We consider the sequence $\{C_s, s = 1, 2, \ldots\}$ of continuous, positive, independent, and identically distributed random variables. The random variables of the sequence $\{C_s, s = 1, 2, \ldots\}$ are equally distributed with the random variable C having characteristic function $\varphi_C(\xi)$.

We suppose that the random variable C_s denotes the size of the sth damage due to the first retained risk occurrence at the time point

$$T = X_1 + X_2 + \cdots + X_N.$$

Hence the random variable

$$L = C_1 + C_2 + \cdots + C_S$$

denotes the total size of the damage due to the first retained risk occurrence at the time point

$$T = X_1 + X_2 + \cdots + X_N.$$

The random variable

$$T = X_1 + X_2 + \cdots + X_N,$$

the random variable W, the random variable

$$V = (X_1 + X_2 + \cdots + X_N)W,$$

the random variable

$$L = C_1 + C_2 + \cdots + C_S$$

and the vector (V, L) are strong analytical tools for investigating and applying risk frequency reduction operations.

The independence of N, $\{X_n, n = 1, 2, \ldots\}$, W, S and $\{C_s, s = 1, 2, \ldots\}$ implies the evaluation of the characteristic function

$$\varphi_{V,L}(u,\xi) = \int_0^\infty \frac{p\varphi_X(uw)}{1-q\varphi_X(uw)} f_W(w)dw \cdot P_S(\varphi_C(\xi))$$

of the vector (V, L).

That characteristic function supports the practical applications in risk management of the vector (V, L).

The above form of the characteristic function $\varphi_{V,L}(u,\xi)$ establishes the independence of the random variable

$$V = (X_1 + X_2 + \cdots + X_N)W$$

and the random variable

$$L = C_1 + C_2 + \cdots + C_s.$$

Hence the vector (V, L) is of particular interest in risk management. □

3.16 Convoluting Cost of Deleting Risk Occurrences with Constant Probability and Total Severity of First Retained Risk Occurrence

Let N be a discrete random variable with values in the set $\mathbf{N} = \{1, 2, \ldots\}$ and following the geometric type II distribution with probability generating function

$$P_N(z) = \frac{pz}{1-qz}.$$

We suppose that $\{X_n, n = 1, 2, \ldots\}$ is a sequence of continuous, positive, and independent random variables. The random variables of the sequence are equally distributed with the random variable X having characteristic function $\varphi_X(u)$ and we set

$$T = X_1 + X_2 + \cdots + X_N.$$

Let W be a continuous and positive random variable with probability density function $f_W(w), w > 0$ and we set

$$V = (X_1 + X_2 + \cdots + X_N)W$$

or equivalently we set $V = TW$.

Let S be a discrete random variable with values in the set $\mathbf{N}_0 = \{0, 1, 2, \ldots\}$ and probability generating function $P_S(z)$.

We suppose that $\{C_s, s = 1, 2, \ldots\}$ is a sequence of continuous, positive, independent, and identically distributed random variables. The random variables of the sequence are equally distributed with the random variable C having characteristic function $\varphi_C(u)$ and we set

$$L = C_1 + C_2 + \cdots + C_S.$$

We consider the random variable

$$Y = (X_1 + X_2 + \cdots + X_N)W + C_1 + C_2 + \cdots + C_S$$

or equivalently the random variable $Y = V + L$.

The purpose of the present section is the establishment of properties and applications in risk management of the random variable $Y = V + L$.

The following result establishes sufficient conditions for evaluating the characteristic function $\varphi_Y(u)$ of the random variable $Y = V + L$.

Theorem 3.16.1 *Let N be a discrete random variable with values in the set $\mathbf{N} = \{1, 2, \ldots\}$ following the geometric type II distribution with probability generating function*

$$P_N(z) = \frac{pz}{1 - qz}.$$

We suppose that $\{X_n, n = 1, 2, \ldots\}$ is a sequence of continuous, positive, independent, and identically distributed random variables. The random variables of the sequence are equally distributed with the random variable X having characteristic function $\varphi_X(u)$ and we set

$$T = X_1 + X_2 + \cdots + X_N.$$

Let W be a continuous and positive random variable with probability density function $f_W(w)$, $w > 0$ and we set

$$V = (X_1 + X_2 + \cdots + X_N)W$$

or equivalently we set $V = TW$.

Let S be a discrete random variable with values in the set $\mathbf{N}_0 = \{0, 1, 2, \ldots\}$ and probability generating function $P_S(z)$.

We suppose that $\{C_s, s = 1, 2, \ldots\}$ is a sequence of continuous, positive, independent, and identically distributed random variables. The random variables of the above sequence are equally distributed with the random variable C having characteristic function $\varphi_C(u)$ and we set

$$L = C_1 + C_2 + \cdots + C_S.$$

We consider the random variable

$$Y = (X_1 + X_2 + \cdots + X_N)W + C_1 + C_2 + \cdots + C_S$$

or equivalently the random variable $Y = V + L$. *If* N, $\{X_n, n = 1, 2, \ldots\}$, W, S *and* $\{C_s, s = 1, 2, \ldots\}$ *are independent then the characteristic function of the random variable* $Y = V + T$ *is*

$$\varphi_Y(u) = \int_0^\infty \frac{p\varphi_X(uw)}{1 - q\varphi_X(uw)} f_W(w)dw \cdot P_S(\varphi_C(u)).$$

Proof From Theorem 3.15.1 it follows that the random variables

$$V = (X_1 + X_2 + \cdots + X_N)W$$

and

$$L = C_1 + C_2 + \cdots + C_S$$

are independent, the random variable

$$V = (X_1 + X_2 + \cdots + X_N)W$$

has characteristic function

$$\varphi_V(u) = \int_0^\infty \frac{p\varphi_X(uw)}{1 - q\varphi_X(uw)} f_W(w)dw$$

and the random variable

$$L = C_1 + C_2 + \cdots + C_S$$

has characteristic function

$$\varphi_L(u) = P_S(\varphi_C(u)).$$

If $\varphi_Y(u)$ is the characteristic function of the random variable $Y = V + L$ then

$$\varphi_Y(u) = E\left(e^{iu(V+L)}\right)$$

or equivalently

$$\varphi_Y(u) = E\left(e^{iuV}\right)E\left(e^{iuL}\right).$$

Hence the characteristic function of the random variable $Y = V + L$ is

$$\varphi_Y(u) = \int_0^\infty \frac{p\varphi_X(uw)}{1 - q\varphi_X(uw)} f_W(w)dw \cdot P_S(\varphi_C(u)).$$

An interpretation of the stochastic model $Y = V + L$ where

$$V = (X_1 + X_2 + \cdots + X_N)W$$

and

$$L = C_1 + C_2 + \cdots + C_S,$$

in the discipline of risk management is the following.

We consider an organization facing a risk at the time point 0. We suppose that $\{X_n, n = 1, 2, \ldots\}$ is a sequence of continuous, positive, independent, and identically distributed random variables. The random variables of the above sequence are equally distributed with the random variable X having characteristic function $\varphi_X(u)$.

The random variable $X_n, n = 1, 2, \ldots$ denotes the time between the $(n-1)$th and the nth risk occurrence. A risk frequency reduction operation is applied to the risk. According to that operation, a risk occurrence is retained with probability p and deleted with probability $q = 1 - p$.

The retention–deletion of a risk occurrence is independent of the retention–deletion of any other risk occurrence. Let N be a random variable denoting the number of risk occurrences until the first retained risk occurrence. The random variable N follows the geometric type II distribution with probability generating function

$$P_N(z) = \frac{pz}{1 - qz}.$$

The sequence of continuous, positive, independent, and identically distributed random variables $\{X_n, n = 1, 2, \ldots\}$ is independent of the random variable N.

The random sum

$$T = X_1 + X_2 + \cdots + X_N$$

denotes the time of the first retained risk occurrence.

Let W be a continuous and positive random variable with probability density function $f_W(w)$. We suppose that the random variable W denotes the cost per unit of time for applying the risk frequency reduction operation. Hence the random variable

$$V = (X_1 + X_2 + \cdots + X_N)W$$

denotes the cost of applying the risk frequency reduction operation until the first retained risk occurrence. We suppose that the random variable S denotes the number of different damages due to the first retained risk occurrence at the time point

$$T = X_1 + X_2 + \cdots + X_N.$$

The random variable S has probability generating function $P_S(z)$.

We consider the sequence $\{C_s, s = 1, 2, \ldots\}$ of continuous, positive, independent, and identically distributed random variables. The random variables of the sequence $\{C_s, s = 1, 2, \ldots\}$ are equally distributed with the random variable C having characteristic function $\varphi_C(u)$.

We suppose that the random variable C_S denotes the size of the sth damage due to the first retained risk occurrence at the time point

$$T = X_1 + X_2 + \cdots + X_N.$$

Hence the random variable

$$L = C_1 + C_2 + \cdots + C_s$$

denotes the total size of the damage due to the first retained risk occurrence at the time point

$$T = X_1 + X_2 + \cdots + X_N.$$

The random variable

$$T = X_1 + X_2 + \cdots + X_N,$$

the random variable W, the random variable

$$V = (X_1 + X_2 + \cdots + X_N)W,$$

the random variable

$$L = C_1 + C_2 + \cdots + C_S,$$

and the random variable $Y = V + L$ are strong analytical tools for investigating and applying risk frequency reduction operations.

The independence of N, $\{X_n, n = 1, 2, \ldots\}$, W, S and $\{C_s, s = 1, 2, \ldots\}$ implies the evaluation of the characteristic function

$$\varphi_Y(u) = \int_0^\infty \frac{p\varphi_X(uw)}{1 - q\varphi_X(uw)} f_W(w)dw \cdot P_S(\varphi_C(u))$$

of the random variable $Y = V + L$, where

$$V = (X_1 + X_2 + \cdots + X_N)W$$

and

$$L = C_1 + C_2 + \cdots + C_S$$

The presence of the random sums

$$T = X_1 + X_2 + \cdots + X_N$$

and

$$L = C_1 + C_2 + \cdots + C_S$$

in the stochastic model $Y = V + L$ supports the practical applications in the discipline of risk management of that model. □

Chapter 4
Stochastic Discounting Modeling for Concepts and Operations of Risk Management

Abstract This chapter concentrates on the formulation, investigation, and applications in real world situations of stochastic discounting models describing fundamental concepts and operations of risk management. The incorporation of important concepts of probability theory in stochastic discounting models makes such stochastic models powerful analytical tools for strategic thinking and strategic decision making. More precisely, the presence of a sum, minimum, and maximum of a random number of continuous, positive, independent, and identically distributed random variables in the mathematical structure of a stochastic model substantially supports the applicability of such a stochastic model in describing, analyzing, selecting, and implementing fundamental risk management operations. In addition, the extremely strong results of the theory of characteristic functions facilitate the use of stochastic discounting models in risk management operations.

4.1 Introduction

The purpose of the present chapter is the formulation, investigation, and practical applications of stochastic discounting models for the fundamental concepts and operations of risk management. More precisely, the chapter concentrates on stochastic models of present value of simple cash flows and stochastic models of present value of continuous and uniform cash flows. These stochastic models substantially support the practical applications of the fundamental concepts and operations of risk management. The concept of sum of a random number of continuous, positive, independent, and identically distributed random variables, the concept of minimum of a random number of continuous, positive, independent, and identically distributed random variables, and the concept of maximum of a random number of continuous, positive, independent, and identically distributed random variables are the structural elements of the stochastic discounting models formulated, investigated, and interpreted by the sections of the present chapter. The presence of the above very important concepts of probability theory in stochastic

C. Artikis and P. Artikis, *Probability Distributions in Risk Management Operations*,
Intelligent Systems Reference Library 83, DOI 10.1007/978-3-319-14256-2_4

discounting models of risk management operations makes these models very strong analytical tools in decision making under conditions of risk. The introduction of the concepts of sum, minimum, and maximum of a random number of continuous, positive, independent, and identically distributed random variables in stochastic discounting models of simple and continuous cash flows constitutes the first part of the theoretical contribution of this chapter. The establishment of sufficient conditions for evaluating the probability distribution of such stochastic models constitutes the second part of the theoretical contribution of this chapter. The practical contribution of this chapter consists of establishing applications of such stochastic models in risk management. From a theoretical and practical point of view the chapter extends the interest in formulating and investigating stochastic discounting models for simple and continuous cash flows in risk management. More precisely, this chapter proposes and investigates stochastic discounting models of simple and continuous cash flows which make clear the theoretical and practical importance of the fundamental concepts and operations of risk management. The recognition of the theoretical and practical significance of the fundamental concepts and operations of risk management is based on the concepts of sum, minimum, and maximum of a random number of continuous, positive, independent, and identically distributed random variables. These important concepts of probability theory are particularly useful in formulating stochastic discounting models incorporating concepts and operations of risk management.

4.2 Present Value of a Single Cash Flow and Proactive Risk Management Decision Making

Let X be a continuous and positive random variable with characteristic function $\varphi_X(u)$ and V be a continuous and positive random variable with characteristic function $\varphi_V(u)$.

We suppose that T is a continuous and positive random variable with distribution function $F_T(t)$ and r is a positive real number.

We suppose that N is a discrete random variable following the Bernoulli distribution with probability function

$$P(N = n) = p^n q^{1-n}, \quad n = 0, 1.$$

We consider the stochastic model

$$Y = \begin{cases} X, & N = 1 \\ Ve^{-rT}, & N = 0. \end{cases}$$

The purpose of the present section is the establishment of properties and applications in risk management of the above stochastic model.

An interpretation of the model

$$Y = \begin{cases} X, & N = 1 \\ Ve^{-rT}, & N = 0 \end{cases}$$

in the area of stochastic discounting of single cash flows is the following.

We suppose that the random variable X denotes a cash flow at the time point 0 and the random variable V denotes a cash flow at the time point T.

The positive real number r denotes the force of interest and the random variable Ve^{-rT} denotes the present value at the time point 0 of the cash flow V corresponding at the time point T.

Hence the stochastic model

$$Y = \begin{cases} X, & N = 1 \\ Ve^{-rT}, & N = 0 \end{cases}$$

denotes the Bernoulli random sum of the above present values at the time point 0.

The following result establishes conditions for evaluating the characteristic function $\varphi_Y(u)$ of the stochastic model

$$Y = \begin{cases} X, & N = 1 \\ Ve^{-rT}, & N = 0. \end{cases}$$

Theorem 4.2.1 *Let X be a continuous and positive random variable with characteristic function $\varphi_X(u)$ and V be a continuous and positive random variable with characteristic function $\varphi_V(u)$.*

We suppose that T is a continuous and positive random variable with distribution function $F_T(t)$, r is a positive real number, and N is a discrete random variable following the Bernoulli distribution with probability function

$$P(N = n) = p^n q^{1-n}, \quad n = 0, 1.$$

If the random variables X, V, T, N are independent then the characteristic function of the stochastic model

$$Y = \begin{cases} X, & N = 1 \\ Ve^{-rT}, & N = 0 \end{cases}$$

is

$$\varphi_Y(u) = p\varphi_X(u) + q \int_0^1 \varphi_V(uw)\, d\left[1 - F_T\left(-\frac{1}{r}\log w\right)\right].$$

Proof The independence of the random variables X, V, T, N implies the independence of the random variables V, T.

We consider the random variable $W = e^{-rT}$.

The independence of the random variables V, T implies the independence of the random variables V, W.

If $F_T(t)$ is the distribution function of the random variable T then

$$F_W(w) = 1 - F_T\left(-\frac{1}{r}\log w\right), \quad 0<w<1, \tag{4.2.1}$$

is the distribution function of the random variable $W = e^{-rT}$.

If $\varphi_{VW}(u)$ is the characteristic function of the random variable VW then we get that

$$\varphi_{VW}(u) = E\left(e^{iuVW}\right)$$

or equivalently

$$\varphi_{VW}(u) = E\left(E\left(e^{iuVW}|W\right)\right). \tag{4.2.2}$$

From (4.2.2) it follows that

$$\varphi_{VW}(u) = \int_0^1 E\left(e^{iuVW}|W = w\right)\, dF_W(w)$$

or equivalently

$$\varphi_{VW}(u) = \int_0^1 E\left(e^{iuwV}|W = w\right)\, dF_W(w). \tag{4.2.3}$$

From (4.2.1) and (4.2.3) it follows that

$$\varphi_{VW}(u) = \int_0^1 E\left(e^{iuwV}|W = w\right) d\left[1 - F_T\left(-\frac{1}{r}\log w\right)\right]. \tag{4.2.4}$$

Since the random variable V is independent of the random variable W then (4.2.4) has the form

$$\varphi_{VW}(u) = \int_0^1 E\left(e^{iuwV}\right) d\left[1 - F_T\left(-\frac{1}{r}\log w\right)\right]. \tag{4.2.5}$$

Since $\varphi_V(u)$ is the characteristic function of the random variable V then (4.2.5) implies that the characteristic function of the random variable VW is

$$\varphi_{VW}(u) = \int_0^1 \varphi_V(uw)\, d\left[1 - F_T\left(-\frac{1}{r}\log w\right)\right]. \tag{4.2.6}$$

Let $\varphi_X(u)$ be the characteristic function of the random variable X, $\varphi_{VW}(\xi)$ the characteristic function of the random variable VW, $\varphi_N(\theta)$ the characteristic function of the random variable N and $\varphi_{X,VW,N}(u,\xi,\theta)$ the characteristic function of the vector (X, VW, N).

The establishment of the relationship

$$\varphi_{X,VW,N}(u,\xi,\theta) = \varphi_X(u)\varphi_{VW}(\xi)\varphi_N(\theta)$$

implies the independence of the random variables X, VW, N.

We have

$$\varphi_{X,VW,N}(u,\xi,\theta) = E\left(e^{iuX+i\xi VW+i\theta N}\right)$$

or equivalently

$$\varphi_{X,VW,N}(u,\xi,\theta) = E\left(E\left(e^{iuX+i\xi VW+i\theta N}|W\right)\right). \tag{4.2.7}$$

From (4.2.7) it follows that

$$\varphi_{X,VW,N}(u,\xi,\theta) = \int_0^1 E\left(e^{iuX+i\xi VW+i\theta N}|W = w\right)\, dF_w(w). \tag{4.2.8}$$

Hence (4.2.1) and (4.2.8) imply that

$$\varphi_{X,VW,N}(u,\xi,\theta) = \int_0^1 E\left(e^{iuX+i\xi wV+i\theta N}|W = w\right)\, d\left[1 - F_T\left(-\frac{1}{r}\log w\right)\right]. \tag{4.2.9}$$

Since the random variables X, V, T, N are independent then the random variables X, V, W, N are also independent. The independence of the above random variables implies the independence of the random variables X, wV, W, N.

Hence (4.2.9) has the form

$$\varphi_{X,VW,N}(u,\xi,\theta) = \int_0^1 E\left(e^{iuX+i\xi wV+i\theta N}\right)\, d\left[1 - F_T\left(-\frac{1}{r}\log w\right)\right]. \tag{4.2.10}$$

The independence of the random variables X, wV, W, N implies the independence of the random variables e^{iuX}, $e^{i\xi wV}$, $e^{i\theta N}$.

Hence (4.2.10) has the form

$$\varphi_{X,VW,N}(u,\xi,\theta) = E\left(e^{iuX}\right)E\left(e^{i\theta N}\right)\int_0^1 E\left(e^{i\xi wV}\right) d\left[1 - F_T\left(-\frac{1}{r}\log w\right)\right]. \quad (4.2.11)$$

From (4.2.6) and (4.2.11) it follows that

$$\varphi_{X,VW,N}(u,\xi,\theta) = \varphi_X(u)\varphi_{VW}(\xi)\varphi_N(\theta). \quad (4.2.12)$$

Moreover (4.2.12) implies that the random variables X, WV, N are independent. If $\varphi_Y(u)$ is the characteristic function of the stochastic model

$$Y = \begin{cases} X, & N = 1 \\ Ve^{-rT}, & N = 0 \end{cases}$$

then we get

$$\varphi_Y(u) = E\left(e^{iuY}\right)$$

or equivalently

$$\varphi_Y(u) = E\left(E\left(e^{iuY}|N\right)\right). \quad (4.2.13)$$

From (4.2.13) it follows that

$$\varphi_Y(u) = \sum_{n=0}^{1} E\left(e^{iuY}|N = n\right)P(N = n). \quad (4.2.14)$$

Hence (4.2.14) implies that

$$\varphi_Y(u) = pE\left(e^{iuY}|N = 1\right) + qE\left(e^{iuY}|N = 0\right). \quad (4.2.15)$$

From (4.2.15) we get that

$$\varphi_Y(u) = pE\left(e^{iuX}|N = 1\right) + qE\left(e^{iuVe^{-rT}}|N = 0\right). \quad (4.2.16)$$

The independence of the random variables

$$X, Ve^{-rT} = VW, N$$

implies the independence of the random variables X, N and the independence of the random variables

$$Ve^{-rT} = VW, N.$$

Hence (4.2.16) has the form

$$\varphi_Y(u) = pE\left(e^{iuX}\right) + qE\left(e^{iuVe^{-rT}}\right). \quad (4.2.17)$$

Since $\varphi_X(u)$ is the characteristic function of the random variable X and

$$\varphi_{VW}(u) = \int_0^1 \varphi_V(uw)\, d\left[1 - F_T\left(-\frac{1}{r}\log w\right)\right]$$

is the characteristic function of the random variable

$$VW = Ve^{-rT}$$

then (4.2.17) implies that the characteristic function of the stochastic model

$$Y = \begin{cases} X, & N = 1 \\ Ve^{-rT}, & N = 0 \end{cases}$$

is

$$\varphi_Y(u) = p\varphi_X(u) + q\int_0^1 \varphi_V(uw)\, d\left[1 - F_T\left(-\frac{1}{r}\log w\right)\right].$$

The evaluation of the characteristic function $\varphi_Y(u)$ substantially supports the practical applications of the corresponding stochastic model in risk management.

An interpretation of the stochastic model

$$Y = \begin{cases} X, & N = 1 \\ Ve^{-rT}, & N = 0 \end{cases}$$

in risk management is the following.

We consider an organization facing a risk at the time point 0. The first choice of the organization is the avoidance of risk by paying an amount X at the time point 0. The second choice of the organization is the retention of risk. The second choice of the organization implies that the organization will pay an amount V, which denotes the severity of risk at the time point T of occurrence of that risk. Hence the stochastic model

$$Y = \begin{cases} X, & N = 1 \\ Ve^{-rT}, & N = 0 \end{cases}$$

denotes the present value at the time point 0 of the cost of the choice for the treatment of that risk.

The following result concentrates on the investigation of the characteristic function $\varphi_Y(u)$ of a special case of the stochastic model

$$Y = \begin{cases} X, & N = 1 \\ Ve^{-rT}, & N = 0. \end{cases}$$

□

Theorem 4.2.2 *Let X be a continuous and positive random variable with differentiable characteristic function $\varphi_X(u)$ and V be a continuous and positive random variable with differentiable characteristic function $\varphi_V(u)$.*

We suppose that T is a continuous and positive random variable following the exponential distribution with distribution function

$$F_T(t) = 1 - e^{-\mu t}, \quad t > 0, \ \mu > 0,$$

and r is a positive real number.

We suppose that N is a discrete random variable following the Bernoulli distribution with probability function

$$P(N = n) = p^n q^{1-n}, \quad n = 0, 1.$$

If X, V, T, N are independent then the characteristic function of the stochastic model

$$Y = \begin{cases} X, & N = 1 \\ Ve^{-rT}, & N = 0 \end{cases}$$

is

$$\varphi_Y(u) = p\varphi_X(u) + qap \int_0^1 \varphi_X(uw) w^{ap-1} dw,$$

where $a = \mu/r$ if, and only, if the random variables Y, V are equally distributed.

Proof From Theorem 4.2.1 it follows that the characteristic function of the stochastic model

$$Y = \begin{cases} X, & N = 1 \\ Ve^{-rT}, & N = 0 \end{cases}$$

is

$$\varphi_Y(u) = p\varphi_X(u) + q \int_0^1 \varphi_V(uw)\, d\left[1 - F_T\left(-\frac{1}{r}\log w\right)\right]. \tag{4.2.18}$$

Since the distribution function of the random variable T is

$$F_T(t) = 1 - e^{-\mu t}, \quad t > 0, \quad \mu > 0,$$

then the distribution function of the random variable $W = e^{-rT}$ is

$$F_W(w) = w^a, \quad 0 < w < 1,$$

where $a = \mu/r$.

Hence (4.2.18) has the form

$$\varphi_Y(u) = p\varphi_X(u) + qa \int_0^1 \varphi_V(uw) w^{a-1} dw$$

or equivalently the form

$$\varphi_Y(u) = p\varphi_X(u) + q\frac{a}{u^a} \int_0^u \varphi_V(w) w^{a-1} dw. \tag{4.2.19}$$

If the random variables Y, V are equally distributed then we get that

$$\varphi_Y(u) = \varphi_V(u). \tag{4.2.20}$$

From (4.2.19) and (4.2.20) it follows that

$$\varphi_Y(u) = p\varphi_X(u) + q\frac{a}{u^a} \int_0^u \varphi_Y(w) w^{a-1} dw. \tag{4.2.21}$$

If we multiply both sides of the integral equation (4.2.21) by u^a then we get the integral equation

$$u^a \varphi_Y(u) = pu^a \varphi_X(u) + qa \int_0^u \varphi_Y(w) w^{a-1} dw. \tag{4.2.22}$$

The characteristic functions $\varphi_Y(u)$ and $\varphi_X(u)$ are differentiable. If we differentiate the integral equation (4.2.22) we get the differential equation

$$au^{a-1}\varphi_Y(u) + u^a\frac{d\varphi_Y(u)}{du} = apu^{a-1}\varphi_X(u) + pu^a\frac{d\varphi_X(u)}{du} + qa\varphi_Y(u)u^{a-1}. \tag{4.2.23}$$

If $u \neq 0$ then the differential equation (4.2.23) implies that

$$\varphi_Y(u) + \frac{u}{ap}\cdot\frac{d\varphi_Y(u)}{du} = \varphi_X(u) + \frac{u}{a}\cdot\frac{d\varphi_X(u)}{du}. \tag{4.2.24}$$

The differential equation (4.2.24) implies the differential equation

$$\left(\varphi_Y(u) + \frac{u}{ap}\cdot\frac{d\varphi_Y(u)}{du}\right)apu^{ap-1} = ap\varphi_X(u)u^{ap-1} + pu^{ap}\frac{d\varphi_X(u)}{du}. \tag{4.2.25}$$

From (4.2.25) it follows that

$$\int_0^u\left(\varphi_Y(y) + \frac{y}{ap}\cdot\frac{d\varphi_Y(y)}{dy}\right)apy^{ap-1}dy = ap\int_0^u \varphi_X(y)y^{ap-1}dy + p\int_0^u y^{ap}d\varphi_X(y). \tag{4.2.26}$$

Hence (4.2.26) implies that

$$ap\int_0^u \varphi_Y(y)y^{ap-1}dy + \int_0^u y^{ap}d\varphi_Y(y) = ap\int_0^u \varphi_X(y)y^{ap-1}dy + p\int_0^u y^{ap}d\varphi_X(y). \tag{4.2.27}$$

From (4.2.27) it follows that

$$\begin{aligned} &ap\int_0^u \varphi_Y(y)y^{ap-1}dy + u^{ap}\varphi_Y(u) - ap\int_0^u \varphi_Y(y)y^{ap-1}dy = \\ &ap\int_0^u \varphi_X(y)y^{ap-1}dy + pu^{ap}\varphi_X(u) - ap^2\int_0^u \varphi_X(y)y^{ap-1}dy. \end{aligned} \tag{4.2.28}$$

Hence (4.2.28) implies that

$$u^{ap}\varphi_Y(u) = pu^{ap}\varphi_X(u) + ap(1-p)\int_0^u \varphi_X(y)y^{ap-1}dy.$$

or equivalently

$$u^{ap}\varphi_Y(u) = pu^{ap}\varphi_X(u) + qap\int_0^u \varphi_X(y)y^{ap-1}dy \tag{4.2.29}$$

If $u \neq 0$ then (4.2.29) implies that

$$\varphi_Y(u) = p\varphi_X(u) + q\frac{ap}{u^{ap}}\int_0^u \varphi_X(y)y^{ap-1}dy$$

or equivalently

$$\varphi_Y(u) = p\varphi_X(u) + qap\int_0^1 \varphi_X(uw)w^{ap-1}dw.$$

The inverse is obvious.
The characteristic function

$$\varphi_Y(u) = p\varphi_X(u) + qap\int_0^1 \varphi_X(uw)w^{ap-1}dw$$

is a mixture of the characteristic function $\varphi_X(u)$ and the characteristic function

$$ap\int_0^1 \varphi_X(uw)w^{ap-1}dw$$

which belongs to a *ap*-unimodal distribution. If the characteristic function $\varphi_X(u)$ belongs to a *ap*-unimodal distribution then it easily follows that the characteristic function

$$\varphi_Y(u) = p\varphi_X(u) + qap\int_0^1 \varphi_X(uw)w^{ap-1}dw$$

or equivalently the characteristic function

$$\varphi_Y(u) = p\varphi_X(u) + q\frac{ap}{u^{ap}}\int_0^u \varphi_X(y)y^{ap-1}dy$$

belongs to *ap*-unimodal distribution. □

4.3 Present Value of Total Risk Severity

Let V be a continuous and positive random variable with characteristic function $\varphi_V(u)$.

We suppose that $\{C_n, n = 1, 2, \ldots\}$ is a sequence of continuous, positive, independent, and identically distributed random variables. The random variables of the sequence are equally distributed with the random variable C having characteristic function $\varphi_C(u)$ and N is a discrete random variable with values in the set $\mathbf{N}_0 = \{0, 1, 2, \ldots\}$ and probability generating function $P_N(z)$.

We set $L = C_1 + C_2 + \cdots + C_N$.

Let U be a continuous and positive random variable with characteristic function $\varphi_U(u)$.

We suppose that T is a continuous and positive random variable with distribution function $F_T(t)$ and r is a positive real number.

We consider the stochastic model

$$Y = V + (C_1 + C_2 + \cdots + C_N + U)e^{-rT}.$$

The purpose of the present section is the establishment of properties and applications in risk management of the stochastic model

$$Y = V + (C_1 + C_2 + \cdots + C_N + U)e^{-rT}.$$

An interpretation of the above model in the area of continuous discounting of simple cash flows is the following.

We suppose that the random variable V denotes a cash flow at the time point 0. Moreover, each random variable of the sequence $\{C_n, n = 1, 2, \ldots\}$ denotes a cash flow at the time point T.

In addition, the random variable U denotes a cash flow at the time point T.

We suppose that the positive real number r denotes force of interest. In this case the stochastic model

$$Y = V + (C_1 + C_2 + \cdots + C_N + U)e^{-rT}$$

denotes the sum of the cash flow V, corresponding at the time point 0, and the present value at the time point 0 of the cash flow $C_1 + C_2 + \cdots + C_N + U$, corresponding at the time point T.

The following result establishes sufficient conditions for evaluating the characteristic function $\varphi_V(u)$. of the stochastic model

$$Y = V + (C_1 + C_2 + \cdots + C_N + U)e^{-rT}.$$

Theorem 4.3.1 *Let V be a continuous and positive random variable with characteristic function $\varphi_V(u)$.*

We suppose that $\{C_n, n = 1, 2, \ldots\}$ is a sequence of continuous, positive, independent, and identically distributed random variables. The random variables of the sequence are equally distributed with the random variable C having characteristic function $\varphi_C(u)$ and N is a discrete random variable with values in the set $\mathbf{N}_0 = \{0, 1, 2, \ldots\}$ *and probability generating function $P_N(z)$.*

We set $L = C_1 + C_2 + \cdots + C_N$.

Let U be a continuous and positive random variable with characteristic function $\varphi_U(u)$. We suppose that T is a continuous and positive random variable with distribution function $F_T(t)$ and r is a positive real number.

We consider the stochastic model

$$Y = V + (C_1 + C_2 + \cdots + C_N + U)e^{-rT}.$$

If V, $\{C_n, n = 1, 2, \ldots\}$, N, U and T are independent then the characteristic function of the stochastic model

$$Y = V + (C_1 + C_2 + \cdots + C_N + U)e^{-rT}$$

is

$$\varphi_Y(Y) = \varphi_V(u) \int_0^1 \varphi_U(uw) P_N(\varphi_C(uw)) d\left[1 - F_T\left(-\frac{1}{r}\log w\right)\right].$$

Proof The independence of V, $\{C_n, n = 1, 2, \ldots\}$, N, U and T implies the independence of $\{C_n, n = 1, 2, \ldots\}$ and N.

Hence (2.5.6) implies that the characteristic function of the random sum $L = C_1 + C_2 + \cdots + C_N$ is

$$\varphi_L(u) = P_N(\varphi_C(u)).$$

Let $\varphi_{V,L,U,T}(u, \xi, \theta, \omega)$ be the characteristic function of the vector (V, L, U, T). The establishment of the relationship

$$\varphi_{V,L,U,T}(u, \xi, \theta, \omega) = \varphi_V(u)\varphi_L(\xi)\varphi_U(\vartheta)\varphi_T(\omega)$$

implies the independence of the random variables V, L, U, T.

We have

$$\varphi_{V,L,U,T}(u,\xi,\theta,\omega) = E\left(e^{iuV+i\xi L+i\theta U+i\omega T}\right)$$

or equivalently

$$\varphi_{V,L,U,T}(u,\xi,\theta,\omega) = E\left(E\left(e^{iuV+i\xi L+i\theta U+i\omega T}|N\right)\right). \tag{4.3.1}$$

From (4.3.1) we get the formula

$$\varphi_{V,L,U,T}(u,\xi,\theta,\omega) = \sum_{n=0}^{\infty} E\left(e^{iuV+i\xi L+i\theta U+i\omega T}|N=n\right)P(N=n)$$

or equivalently the formula

$$\varphi_{V,L,U,T}(u,\xi,\theta,\omega) = \sum_{n=0}^{\infty} E\left(e^{iuV+i\xi(C_1+\cdots+C_N)+i\theta U+i\omega T}|N=n\right)P(N=n)$$

which can be written in the form

$$\varphi_{V,L,U,T}(u,\xi,\theta,\omega) = \sum_{n=0}^{\infty} E\left(e^{iuV+i\xi(C_1+\cdots+C_n)+i\theta U+i\omega T}|N=n\right)P(N=n). \tag{4.3.2}$$

Hence (4.3.2) has the form

$$\varphi_{V,L,U,T}(u,\xi,\theta,\omega) = \sum_{n=0}^{\infty} E\left(e^{iuV}e^{i\xi C_1}\ldots e^{i\xi C_n}e^{i\theta U}e^{i\omega T}|N=n\right)P(N=n). \tag{4.3.3}$$

The independence of V, $\{C_n, n=1,2,\ldots\}$, N, U and T implies the independence of the random variables $V, C_1,\ldots, C_n, N, U, T$.

The independence of the above random variables implies the independence of the random variables $e^{iuV}, e^{i\xi C_1},\ldots, e^{i\xi C_n}, N, e^{i\theta U}, e^{i\omega T}$.

Hence (4.3.3) has the form

$$\varphi_{V,L,U,T}(u,\xi,\theta,\omega) = \sum_{n=0}^{\infty} E\left(e^{iuV}e^{i\xi C_1}\ldots e^{i\xi C_n}e^{i\theta U}e^{i\omega T}\right)P(N=n). \tag{4.3.4}$$

Since the independence of the random variables $e^{iuV}, e^{i\xi C_1},\ldots, e^{i\xi C_n}, N, e^{i\theta U}, e^{i\omega T}$ implies the independence of the random variables $e^{iuV}, e^{i\xi C_1},\ldots, e^{i\xi C_n}, e^{i\theta U}, e^{i\omega T}$ then (4.3.4) has the form

$$\varphi_{V,L,U,T}(u,\xi,\theta,\omega) = E(e^{iuV})E(e^{i\theta U})E(e^{i\omega T})\sum_{n=0}^{\infty}E(e^{i\xi C_1})\ldots E(e^{i\xi C_n})P(N=n). \tag{4.3.5}$$

Since the random variables of the sequence $\{C_n, n=1,2,\ldots\}$ are equally distributed with the random variable C having characteristic function $\varphi_C(u)$ then (4.3.5) has the form

$$\varphi_{V,L,U,T}(u,\xi,\theta,\omega) = \varphi_V(u)P_N(\varphi_C(\xi))\varphi_U(\vartheta)\varphi_T(\omega)$$

or equivalently the form

$$\varphi_{V,L,U,T}(u,\xi,\theta,\omega) = \varphi_V(u)\varphi_L(\xi)\varphi_U(\vartheta)\varphi_T(\omega).$$

Hence the random variables V, L, U, T are independent. Let $\varphi_Y(u)$ be the characteristic function of the stochastic model

$$Y = V + (C_1 + C_2 + \cdots + C_N + U)e^{-rT}$$

and $F_W(w)$ be the distribution function of the random variable $W = e^{-rT}$.
We have

$$F_W(w) = 1 - F_T\left(-\frac{1}{r}\log w\right), \quad 0<w<1. \tag{4.3.6}$$

The independence of the random variables V, L, U, T implies the independence of the random variables V, L, U, W.
Hence

$$\varphi_Y(u) = E\left(e^{iu[V+(C_1+\cdots+C_N+U)W]}\right)$$

or equivalently

$$\varphi_Y(u) = E\left(E\left(e^{iu[V+(C_1+\cdots+C_N+U)W]}|W\right)\right). \tag{4.3.7}$$

From (4.3.7) it follows that

$$\varphi_Y(u) = \int_0^1 E\left(e^{iu[V+(C_1+\cdots+C_N+U)w]}|W=w\right)dF_w(w). \tag{4.3.8}$$

From (4.3.6) and (4.3.8) it follows that

$$\varphi_Y(u) = \int_0^1 E\left(e^{iu[V+(C_1+\cdots+C_N+U)W]}\middle| W = w\right) d\left[1 - F_T\left(-\frac{1}{r}\log w\right)\right]. \quad (4.3.9)$$

Since the random variables V, L, U, W are independent then the random variables V, wL, wU, W are also independent. Hence (4.3.9) has the form

$$\varphi_Y(u) = \int_0^1 E\left(e^{iu[V+(C_1+\cdots+C_N+U)w]}\right) d\left[1 - F_T\left(-\frac{1}{r}\log w\right)\right]. \quad (4.3.10)$$

The independence of the random variables V, wL, wU, W implies the independence of the random variables V, wL, wU.

Hence (4.3.10) has the form

$$\varphi_Y(u) = \int_0^1 E\left(e^{iuV}\right) E\left(e^{iu(C_1+\cdots+C_N)w}\right) E\left(e^{iuwU}\right) d\left[1 - F_T\left(-\frac{1}{r}\log w\right)\right]. \quad (4.3.11)$$

From (4.3.11) it follows that the characteristic function of the stochastic model

$$Y = V + (C_1 + C_2 + \cdots + C_N + U)e^{-rT}$$

is

$$\varphi_Y(u) = \varphi_V(u) \int_0^1 \varphi_U(uw) P_N(\varphi_C(uw)) d\left[1 - F_T\left(-\frac{1}{r}\log w\right)\right].$$

An interpretation of the stochastic model

$$Y = V + (C_1 + C_2 + \cdots + C_N + U)e^{-rT}$$

in risk management is the following.

We suppose that a risk occurs at the time point 0. The random variable V denotes the size of the damage due to the occurrence of risk. Moreover, we suppose that another risk occurs at the time point T.

The continuous and positive random U denotes the size of the damage due to the risk occurrence at the time point T.

In addition, the discrete random variable N denotes the number of different damages due to the risk occurrence at the above random time point and the random variable X_n denotes the size of the nth damage. Hence the random variable

$$C_1 + C_2 + \cdots + C_N + U$$

denotes the total size of the damage due to the risk occurring at the time point T and the random variable

$$(C_1 + C_2 + \cdots + C_N + U)e^{-rT}$$

denotes the present value at the time point 0 of that total damage. The stochastic model

$$Y = V + (C_1 + C_2 + \cdots + C_N + U)e^{-rT}$$

denotes the sum of the damage due to the risk occurring at the time point 0 and the present value of the damage due the risk occurring at the time point T.

The following result concentrates on the investigation of the characteristic function $\varphi_Y(u)$ which corresponds to a special case of the stochastic model

$$Y = V + (C_1 + C_2 + \cdots + C_N + U)e^{-rT}.$$

□

Theorem 4.3.2 *Let V be a continuous and positive random variable with characteristic function $\varphi_V(u)$ and finite mean value.*

We suppose that $\{C_n, n = 1, 2, \ldots\}$ is a sequence of continuous, positive, independent, and identically distributed random variables. The random variables of the sequence are equally distributed with the random variable C having characteristic function $\varphi_C(u)$ and finite mean. Let N be a discrete random variable with values in the set $\mathbf{N}_0 = \{0, 1, 2, \ldots\}$, probability generating function $P_N(z)$ and finite mean. We set $L = C_1 + C_2 + \cdots + C_N$.

Let U be a continuous and positive random variable with characteristic function $\varphi_U(u)$ and finite mean. We suppose that T is a continuous and positive random variable following the exponential distribution with distribution function

$$F_T(t) = 1 - e^{-\mu t}, \quad t > 0 \;\; \mu > 0$$

and r is positive real number.

We consider the stochastic model

$$Y = V + (C_1 + C_2 + \cdots + C_N + U)e^{-rT}.$$

If V, $\{C_n,\; n = 1, 2, \ldots\}$, N, U and T are independent then the characteristic function of the stochastic model

$$Y = V + (C_1 + C_2 + \cdots + C_N + U)e^{-rT}$$

is

$$\varphi_Y(u) = \varphi_V(u) \exp\left(a \int_0^u \frac{\varphi_V(y) P_N(\varphi_C(y)) - 1}{y} dy \right),$$

where $a = \mu/r$, *if, and only, if the random variables* Y, U *are equally distributed.*

Proof From Theorem 4.3.1 it follows that the characteristic function of the stochastic model

$$Y = V + (C_1 + C_2 + \cdots + C_N + U)e^{-rT}$$

is

$$\varphi_Y(u) = \varphi_V(u) \int_0^1 \varphi_U(uw) P_N(\varphi_C(uw)) d\left[1 - F_T\left(-\frac{1}{r}\log w\right)\right]. \quad (4.3.12)$$

Since the distribution function of the random variable T is

$$F_T(t) = 1 - e^{-\mu t}, \quad t > 0, \quad \mu > 0.$$

then the distribution function of the random variable $W = e^{-rT}$ is

$$F_W(w) = w^a, \quad 0 < w < 1,$$

where $a = \mu/r$.

Hence (4.3.12) has the form

$$\varphi_Y(u) = \varphi_V(u) a \int_0^1 \varphi_U(uw) P_N(\varphi_C(uw)) w^{a-1} dw$$

or equivalently the form

$$\varphi_Y(u) = \varphi_V(u) \frac{a}{u^a} \int_0^u \varphi_U(w) P_N(\varphi_C(w)) w^{a-1} dw. \quad (4.3.13)$$

If the random variables Y, U are equally distributed then we get that

$$\varphi_Y(u) = \varphi_U(u). \quad (4.3.14)$$

From (4.3.13) and (4.3.14) it follows that

$$\varphi_Y(u) = \varphi_V(u)\frac{a}{u^a}\int_0^u \varphi_Y(w)P_N(\varphi_C(w))w^{a-1}dw. \tag{4.3.15}$$

If we multiply both sides of the integral equation (4.3.15) by $\frac{u^a}{\varphi_V(u)}$, with u such that $\varphi_V(u) \neq 0$ then we get the integral equation

$$u^a\frac{\varphi_Y(u)}{\varphi_V(u)} = a\int_0^u \varphi_Y(w)P_N(\varphi_C(w))w^{a-1}dw. \tag{4.3.16}$$

If we differentiate the integral equation (4.3.16) we get the differential equation

$$au^{a-1}\frac{\varphi_Y(u)}{\varphi_V(u)} + u^a\frac{1}{\varphi_V(u)}\cdot\frac{d\varphi_Y(u)}{du} - u^a\frac{\varphi_Y(u)}{\varphi_V^2(u)}\cdot\frac{d\varphi_V(u)}{du} = a\varphi_Y(u)P_N(\varphi_C(u))u^{a-1}. \tag{4.3.17}$$

If $u \neq 0$ then the differential equation (4.3.17) has the form

$$a\frac{\varphi_Y(u)}{\varphi_V(u)} + \frac{u}{\varphi_V(u)}\cdot\frac{d\varphi_Y(u)}{du} - u\frac{\varphi_Y(u)}{\varphi_V^2(u)}\cdot\frac{d\varphi_V(u)}{du} = a\varphi_Y(u)P_N(\varphi_C(u)). \tag{4.3.18}$$

Moreover, if we multiply both sides of the differential equation (4.3.18) by $\varphi_V(u)$ then we get the differential equation

$$a\varphi_Y(u) + u\cdot\frac{d\varphi_Y(u)}{du} - u\frac{\varphi_Y(u)}{\varphi_V(u)}\cdot\frac{d\varphi_V(u)}{du} = a\varphi_V(u)\varphi_Y(u)P_N(\varphi_C(u)). \tag{4.3.19}$$

If we divide both sides of the differential equation (4.3.19) by $\varphi_Y(u)$ where u such that $\varphi_Y(u) \neq 0$ then we get the differential equation

$$a + \frac{u}{\varphi_Y(u)}\cdot\frac{d\varphi_Y(u)}{du} - \frac{u}{\varphi_V(u)}\cdot\frac{d\varphi_V(u)}{du} = a\varphi_V(u)P_N(\varphi_C(u)). \tag{4.3.20}$$

From the differential equation (4.3.20) it follows the differential equation

$$\frac{u}{\varphi_Y(u)}\cdot\frac{d\varphi_Y(u)}{du} - \frac{u}{\varphi_V(u)}\cdot\frac{d\varphi_V(u)}{du} = a(\varphi_V(u)P_N(\varphi_C(u)) - 1). \tag{4.3.21}$$

If $u \neq 0$ then the differential equation (4.3.21) has the form

$$\frac{1}{\varphi_Y(u)}\cdot\frac{d\varphi_Y(u)}{du} - \frac{1}{\varphi_V(u)}\cdot\frac{d\varphi_V(u)}{du} = a\frac{\varphi_V(u)P_N(\varphi_C(w)) - 1}{u}. \tag{4.3.22}$$

If we integrate both sides of the differential equation (4.3.22) then we get that

$$\int_0^u \frac{d\varphi_Y(y)}{\varphi_Y(y)} - \int_0^u \frac{d\varphi_V(y)}{\varphi_V(y)} = a \int_0^u \frac{\varphi_V(y)P_N(\varphi_C(y)) - 1}{y} dy. \tag{4.3.23}$$

From (4.3.23) it follows that

$$\log \varphi_Y(u) - \log \varphi_V(u) = a \int_0^u \frac{\varphi_V(y)P_N(\varphi_C(y)) - 1}{y} dy$$

or equivalently

$$\log \varphi_Y(u) - \log \varphi_V(u) = \log \exp\left(a \int_0^u \frac{\varphi_V(y)P_N(\varphi_C(y)) - 1}{y} dy \right). \tag{4.3.24}$$

From (4.3.24) it follows that

$$\log \frac{\varphi_Y(u)}{\varphi_V(u)} = \log \exp\left(a \int_0^u \frac{\varphi_V(y)P_N(\varphi_C(Y)) - 1}{y} dy \right)$$

or equivalently

$$\frac{\varphi_Y(u)}{\varphi_V(u)} = \exp\left(a \int_0^u \frac{\varphi_V(y)P_N(\varphi_C(y)) - 1}{y} dy \right) \tag{4.3.25}$$

with u such that $\varphi_V(u) \neq 0$.

From (4.3.25) it follows that

$$\varphi_Y(u) = \varphi_V(u) \exp\left(a \int_0^u \frac{\varphi_V(y)P_N(\varphi_C(y)) - 1}{y} dy \right).$$

The inverse is obvious.

Since the function

$$\varphi_B(u) = \exp\left(a \int_0^u \frac{\varphi_V(y)P_N(\varphi_C(y)) - 1}{y} dy \right)$$

is the characteristic function of a self decomposable distribution and we suppose that the characteristic function $\varphi_V(u)$ belongs to a selfdecomposable distribution then we conclude that the characteristic function

$$\varphi_Y(u) = \varphi_V(u) \exp\left(a \int_0^u \frac{\varphi_V(y) P_N(\varphi_C(y)) - 1}{y} dy \right)$$

belongs to a selfdecomposable distribution. Hence the probability density function corresponding to the characteristic function

$$\varphi_Y(u) = \varphi_V(u) \exp\left(a \int_0^u \frac{\varphi_V(y) P_N(\varphi_C(y)) - 1}{y} dy \right)$$

is unimodal. □

4.4 Recovery Time of a Partially Damaged System and Present Value of a Single Cash Flow

Let $\{C_s, s = 1, 2, \ldots\}$ a be a sequence of continuous, positive, independent, and identically distributed random variables. We consider the discrete random variable S with values in the set $\mathbf{N}_0 = \{1, 2, \ldots\}$ and probability generating function $P_S(z)$.

We consider the random variable $L = C_1 + C_2 + \cdots + C_S$.

Let $\{X_n, n = 1, 2, \ldots\}$ be a sequence of continuous, positive, independent, and identically distributed random variables. We consider the discrete random variable N with values in the set $\mathbf{N} = \{0, 1, 2, \ldots\}$ and probability generating function $P_N(z)$.

We consider the random variable $T = \max(X_1, X_2, \ldots, X_N)$ and the stochastic model

$$Y = Le^{-rT}$$

where r is a positive real number. The purpose of the present section is the investigation and the applications in risk management of the above stochastic model. An interpretation in the area of continuous discounting of that model is the following.

We suppose that each random variable of the sequence $\{X_n, n = 1, 2, \ldots\}$ denotes time, each random variable of the sequence $\{C_s, s = 1, 2, \ldots\}$ denotes a cash flow and the positive real number r denotes force of interest. In this case, the stochastic model

$$Y = Le^{-rT}$$

denotes the present value at the time point 0 of the cash flow $L = C_1 + C_2 + \cdots + C_S$ corresponding at the time point $T = \max(X_1, X_2, \ldots, X_N)$.

The following result establishes sufficient conditions for evaluating the characteristic function of the stochastic model

$$Y = Le^{-rT}.$$

Theorem 4.4.1 *Let $\{C_s, s = 1, 2, \ldots\}$ be a sequence of continuous, positive, independent, and identically distributed random variables. The random variables of the sequence $\{C_s, s = 1, 2, \ldots\}$ are equally distributed with the random variable C having characteristic function*

$$\varphi_C(u). \tag{4.4.1}$$

We consider the discrete random variable S with values in the set $\mathbf{N}_0 = \{0, 1, 2, \ldots\}$ and probability generating function

$$P_S(z) \tag{4.4.2}$$

and we set $L = C_1 + C_2 + \cdots + C_S$.

Let $\{X_n, n = 1, 2, \ldots\}$ is a sequence of continuous, positive, independent, and identically distributed random variables. The random variables of the sequence $\{X_n, n = 1, 2, \ldots\}$ are equally distributed with the random variable X having distribution function

$$F_X(x). \tag{4.4.3}$$

We consider the discrete random variable N with values in the set $\mathbf{N} = \{1, 2, \ldots\}$ and probability generating function

$$P_N(z) \tag{4.4.4}$$

and we set $T = \max(X_1, X_2, \ldots, X_N)$.

We consider the stochastic model

$$Y = Le^{-rT}$$

where r is a positive real number. If $\{C_s, s = 1, 2, \ldots\}$, S, $\{X_n, n = 1, 2, \ldots\}$ and N are independent then the characteristic function of the stochastic model

$$Y = Le^{-rT}$$

is

$$\varphi_Y(u) = \int_0^1 P_S(\varphi_C(uw))d\left[1 - P_N\left(F_X\left(-\frac{1}{r}\log w\right)\right)\right].$$

Proof We consider the random variable $L = C_1 + C_2 + \cdots + C_S$.

The independence of $\{C_s, s = 1, 2, \ldots\}$, S, $\{X_n, n = 1, 2, \ldots\}$ and N implies the independence of S and $\{C_s, s = 1, 2, \ldots\}$.

From (2.5.6), (4.4.1), and (4.4.2) it follows that the characteristic function of the random variable $L = C_1 + C_2 + \cdots + C_S$ is

$$\varphi_L(u) = P_S(\varphi_C(u)).$$

Let $\varphi_T(\xi)$ be the characteristic function of the random variable $T = \max(X_1, X_2, \ldots, X_N)$ and $\varphi_{L,T}(u, \xi)$ be the characteristic function of the vector (L, T) of random variables $L = C_1 + C_2 + \cdots + C_S$ and $T = \max(X_1, X_2, \ldots, X_N)$.

The independence of the random variable $L = C_1 + C_2 + \cdots + C_S$ and the random variable $T = \max(X_1, X_2, \ldots, X_N)$ is required for evaluating the characteristic function $\varphi_Y(u)$ of the stochastic model

$$Y = Le^{-rT}.$$

The establishment of the relationship

$$\varphi_{L,T}(u, \xi) = \varphi_L(u)\varphi_T(\xi) \tag{4.4.5}$$

implies the independence of the random variables $L = C_1 + C_2 + \cdots + C_S$ and $T = \max(X_1, X_2, \ldots, X_N)$.

We have

$$\varphi_{L,T}(u, \xi) = E\left(e^{iuL + i\xi T}\right)$$

or equivalently we have

$$\varphi_{L,T}(u, \xi) = E\left(E\left(e^{iuL + i\xi T}|S\right)\right). \tag{4.4.6}$$

From (4.4.6) it follows that

$$\varphi_{L,T}(u,\xi) = \sum_{s=0}^{\infty} E\left(e^{iuL+i\xi T}|S=s\right)P(S=s). \tag{4.4.7}$$

Hence (4.4.7) implies that

$$\varphi_{L,T}(u,\xi) = \sum_{s=0}^{\infty} E\left(e^{iu(C_1+\cdots+C_S)+i\xi T}|S=s\right)P(S=s)$$

or equivalently

$$\varphi_{L,T}(u,\xi) = \sum_{s=0}^{\infty} E\left(e^{iuC_1+\cdots+iuC_s+i\xi T}|S=s\right)P(S=s). \tag{4.4.8}$$

From (4.4.8) it follows that

$$\varphi_{L,T}(u,\xi) = \sum_{s=0}^{\infty} E\left(e^{iuC_1}\ldots e^{iuC_s}e^{i\xi T}|S=s\right)P(S=s). \tag{4.4.9}$$

We shall show that the random variables $S, C_1, \ldots, C_s, T$ are independent. The random variables of the sequence $\{C_s, s = 1, 2, \ldots\}$ are equally distributed with the random variable C having distribution function $F_C(c)$.

The independence of $\{C_s, s = 1, 2, \ldots\}$, S, $\{X_n, n = 1, 2, \ldots\}$ and N implies the independence of N and $\{X_n, n = 1, 2, \ldots\}$.

Hence (2.6.7), (4.4.3) and (4.4.4) imply that the distribution function $F_T(t)$ of the random variable $T = \max(X_1, X_2, \ldots, X_N)$ is

$$F_T(t) = P_N(F_X(t)). \tag{4.4.10}$$

We consider the vector $(S, C_1, \ldots, C_s, T)$ of random variables $S, C_1, \ldots, C_s, T$ and let $F_{S,C_1,\ldots C_s,T}(s, c_1, \ldots, c_s, t)$ be the distribution function of that vector. Let $F_S(s)$ be the distribution function of the random variable S.

The proof of independence of the random variables $S, C_1, \ldots, C_s, T$ requires the proof of the relationship

$$F_{S,C_1,\ldots C_s,T}(s, c_1, \ldots, c_s, t) = F_S(s)F_{C_1}(c_1)\ldots F_{C_s}(c_s)F_T(t)$$

or equivalently the proof of the relationship

$$F_{S,C_1,\ldots C_s,T}(s, c_1, \ldots, c_s, t) = F_s(s)F_C(c_1)\ldots F_C(c_s)F_T(t).$$

We have

$$P(S \le s,\ C_1 \le c_1, \ldots, C_s \le c_s) = P(S \le s,\ C_1 \le c_1, \ldots, C_s \le c_s, T \le t) + P(S \le s,\ C_1 \le c_1, \ldots, C_s \le c_s, T > t). \tag{4.4.11}$$

From (4.4.11) it follows that

$$P(S \le s,\ C_1 \le c_1, \ldots, C_s \le c_s, T > t) = P(S \le s,\ C_1 \le c_1, \ldots, C_s \le c_s) - P(S \le s,\ C_1 \le c_1, \ldots, C_s \le c_s, T \le t). \tag{4.4.12}$$

Since the random variable S is independent of the sequence of random variables $\{C_s, s = 1, 2, \ldots\}$ then the random variables S, C_1, …, C_s are independent. Hence

$$P(S \le s,\ C_1 \le c_1, \ldots, C_s \le c_s) = P(S \le s)P(C_1 \le c_1)\ldots P(C_s \le c_s)$$

or equivalently we get

$$P(S \le s,\ C_1 \le c_1, \ldots, C_s \le c_s) = F_S(s)F_{C_1}(c_1)\ldots F_{C_s}(c_s). \tag{4.4.13}$$

From (4.4.13) it follows that

$$P(S \le s,\ C_1 \le c_1, \ldots, C_s \le c_s) = F_S(s)F_C(c_1)\ldots F_C(c_s). \tag{4.4.14}$$

Moreover (4.4.12) and (4.4.14) imply that

$$P(S \le s,\ C_1 \le c_1, \ldots, C_s \le c_s, T > t) = F_S(s)F_C(c_1)\ldots F_C(c_s) - P(S \le s,\ C_1 \le c_1, \ldots, C_s \le c_s, T \le t). \tag{4.4.15}$$

Hence (4.4.15) implies that

$$P(S \le s,\ C_1 \le c_1, \ldots, C_s \le c_s, T > t) = F_S(s)F_C(c_1)\ldots F_C(c_s) - \sum_{n=1}^{\infty} P(S \le s,\ C_1 \le c_1, \ldots, C_s \le c_s, T \le t | N = n)P(N = n). \tag{4.4.16}$$

From (4.4.16) it follows that

$$P(S \leq s, C_1 \leq c_1, \ldots, C_s \leq c_s, T > t) = F_s(s)F_C(c_1)\ldots F_C(c_s) - \sum_{n=1}^{\infty} P[S \leq s, C_1 \leq c_1, \ldots, C_s \leq c_s, \max(X_1, \ldots, X_N) \leq t | N = n]P(N = n). \tag{4.4.17}$$

From (4.4.17) we get that

$$P(S \leq s, C_1 \leq c_1, \ldots, C_s \leq c_s, T > t) = F_s(s)F_C(c_1)\ldots F_C(c_s) - \sum_{n=1}^{\infty} P[S \leq s, C_1 \leq c_1, \ldots, C_s \leq c_s, \max(X_1, \ldots, X_n) \leq t | N = n]P(N = n). \tag{4.4.18}$$

Hence (4.4.18) implies that

$$P(S \leq s, C_1 \leq c_1, \ldots, C_s \leq c_s, T > t) = F_s(s)F_C(c_1)\ldots F_C(c_s) - \sum_{n=1}^{\infty} P(S \leq s, C_1 \leq c_1, \ldots, C_s \leq c_s, X_1 \leq t, \ldots, X_n \leq t | N = n)P(N = n). \tag{4.4.19}$$

Since N, $\{X_{n,} n = 1, 2, \ldots\}$, S, $\{C_{s,} s = 1, 2, \ldots\}$ are independent then the random variables $N, S, C_1, \ldots, C_s, X_1, \ldots, X_n$ are independent. Hence (4.4.19) has the form

$$P(S \leq s, C_1 \leq c_1, \ldots, C_s \leq c_s, T > t) = F_S(s)F_C(c_1)\ldots F_C(c_s) - \sum_{n=1}^{\infty} P(S \leq s, C_1 \leq c_1, \ldots, C_s \leq c_s, X_1 \leq t, \ldots, X_n \leq t)P(N = n). \tag{4.4.20}$$

The independence of the random variables $N, S, C_1, \ldots, C_s, X_1, \ldots, X_n$ implies the independence the independence of the random variables $S, C_1, \ldots, C_s, X_1, \ldots, X_n$.
Hence (4.4.20) has the form

$$P(S \le s, C_1 \le c_1, \ldots, C_s \le c_s, T > t) = F_s(s)F_C(c_1)\ldots F_C(c_s) - \sum_{n=1}^{\infty} P(S \le s)P(C_1 \le c_1)\ldots P(C_s \le c_s) \times P(X_1 \le t)\ldots P(X_n \le t)P(N = n)$$

or equivalently (4.4.20) has the form

$$P(S \le s, C_1 \le c_1, \ldots, C_s \le c_s, T > t) = F_s(s)F_c(c_1)\ldots F_c(c_s) - P(S \le s)P(C_1 \le c_1)\ldots P(C_s \le c_s) \times \sum_{n=1}^{\infty} P(X_1 \le t)\ldots P(X_n \le t)P(N = n). \tag{4.4.21}$$

From (4.4.21) it follows that

$$P(S \le s, C_1 \le c_1, \ldots, C_s \le c_s, T > t) = F_s(s)F_c(c_1)\ldots F_c(c_s) - F_s(s)F_c(c_1)\ldots F_c(c_s)\sum_{n=1}^{\infty} F_X^n(t)P(N = n). \tag{4.4.22}$$

Hence (4.4.22) implies that

$$P(S \le s, C_1 \le c_1, \ldots, C_s \le c_s, T > t) = F_s(s)F_c(c_1)\ldots F_c(c_s) - F_s(s)F_c(c_1)\ldots F_c(c_s)P_N(F_X(t)). \tag{4.4.23}$$

From (4.4.10) and (4.4.23) we get that

$$P(S \le s, C_1 \le c_1, \ldots, C_s \le c_s, T > t) = F_s(s)F_c(c_1)\ldots F_c(c_s) - F_s(s)F_c(c_1)\ldots F_c(c_s)F_T(t)$$

or equivalently we get that

$$P(S \le s, C_1 \le c_1, \ldots, C_s \le c_s, T > t) = F_s(s)F_c(c_1)\ldots F_c(c_s)(1 - F_T(t)). \tag{4.4.24}$$

Since

$$P(S \le s, C_1 \le c_1, \ldots, C_s \le c_s) = P(S \le s, C_1 \le c_1, \ldots, C_s \le c_s, T \le t) + P(S \le s, C_1 \le c_1, \ldots, C_s \le c_s, T > t)$$

then (4.4.24) implies that

$$P(S \leq s, C_1 \leq c_1, \ldots, C_s \leq c_s, T \leq t) = P(S \leq s, C_1 \leq c_1, \ldots, C_s \leq c_s) \\ - F_s(s)F_c(c_1)\ldots F_c(c_s)(1 - F_T(t)). \tag{4.4.25}$$

The independence of the random variables S, C_1, …, C_s implies that

$$P(S \leq s, C_1 \leq c_1, \ldots, C_s \leq c_s) = P(S \leq s)P(C_1 \leq c_1)\ldots P(C_s \leq c_s). \tag{4.4.26}$$

From (4.4.26) we get that

$$P(S \leq s, C_1 \leq c_1, \ldots, C_s \leq c_s) = F_s(s)F_c(c_1)\ldots F_c(c_s) \tag{4.4.27}$$

Moreover (4.4.25) and (4.4.27) imply that

$$P(S \leq s, C_1 \leq c_1, \ldots, C_s \leq c_s, T \leq t) = F_s(s)F_c(c_1)\ldots F_c(c_s) \\ - F_s(s)F_c(c_1)\ldots F_c(c_s)(1 - F_T(t)). \tag{4.4.28}$$

Hence (4.4.28) implies that

$$P(S \leq s, C_1 \leq c_1, \ldots, C_s \leq c_s, T \leq t) = F_s(s)F_c(c_1)\ldots F_c(c_s)F_T(t). \tag{4.4.29}$$

Since

$$P(S \leq s, C_1 \leq c_1, \ldots, C_s \leq c_s, T \leq t) = F_{S,C_1,\ldots,C_s,T}(s, c_1, \ldots, c_s, t)$$

then (4.4.29) implies that

$$F_{S,C_1,\ldots,C_s,T}(s, c_1, \ldots, c_s, t) = F_s(s)F_c(c_1)\ldots F_c(c_s)F_T(t)$$

or equivalently implies that

$$F_{S,C_1,\ldots,C_s,T}(s, c_1, \ldots, c_s, t) = F_s(s)F_{c_1}(c_1)\ldots F_{c_s}(c_s)F_T(t).$$

Hence the random variables $S, C_1, \ldots, C_S, T$ are independent. That implies the independence of the random variables S, e^{iuC_1}, …, e^{iuC_s}, $e^{i\xi T}$.

Hence (4.4.9) has the form

$$\varphi_{L,T}(u, \xi) = \sum_{s=0}^{\infty} E\left(e^{iuC_1}\ldots e^{iuC_s}e^{i\xi T}\right)P(S = s).$$

Since the independence of the random variables S, e^{iuC_1}, …, e^{iuC_s}, $e^{i\xi T}$ implies the independence of the random variables $e^{iuC_1}\ldots e^{iuC_s}e^{i\xi T}$ then we get that

$$\varphi_{L,T}(u,\xi) = \sum_{s=0}^{\infty} E\left(e^{iuC_1}\right)\ldots E\left(e^{iuC_s}\right)E\left(e^{i\xi T}\right)P(S=s). \tag{4.4.30}$$

Since the random variables of the sequence $\{C_s, s=1,2,\ldots\}$ are equally distributed with the random variable C having characteristic function $\varphi_C(u)$ and

$$\varphi_T(\xi) = E\left(e^{i\xi T}\right)$$

is the characteristic function of the random variable $T = \max(X_1, X_2, \ldots, X_N)$ then (4.4.30) has the form

$$\varphi_{L,T}(u,\xi) = \varphi_T(\xi)\sum_{s=0}^{\infty} \varphi_C^s(u)P(S=s). \tag{4.4.31}$$

From (4.4.31) it follows that

$$\varphi_{L,T}(u,\xi) = \varphi_T(\xi)P_S(\varphi_C(u)). \tag{4.4.32}$$

Since

$$\varphi_L(u) = P_S(\varphi_C(u))$$

is the characteristic function of the random sum $L = C_1 + C_2 + \cdots + C_S$ then (4.4.32) has the form

$$\varphi_{L,T}(u,\xi) = \varphi_L(u)\varphi_T(\xi)$$

Hence (4.4.5) implies the independence of the random variables $L = C_1 + C_2 + \cdots + C_S$ and $T = \max(X_1, X_2, \ldots, X_N)$.

We consider the random variable $W = e^{-rT}$.

The independence of the random variables $L = C_1 + C_2 + \cdots + C_S$ and $T = \max(X_1, X_2, \ldots, X_N)$ implies the independence of the random variables $L = C_1 + C_2 + \cdots + C_S$ and $W = \exp[-r\max(X_1, X_2, \ldots, X_N)]$.

The evaluation of the characteristic function $\varphi_Y(u)$ of the stochastic model

$$Y = Le^{-rT}$$

or equivalently $Y = LW$ requires the evaluation of the distribution function $F_W(w)$ of the random variable $W = e^{-rT}$.

From (4.4.10) we get that

$$F_T(t) = P_N(F_X(t))$$

is the distribution function of the random variable $T = \max(X_1, X_2, \ldots, X_N)$.

We have

$$F_W(w) = P(W \le w)$$

or equivalently we have

$$F_W(w) = P(e^{-rT} \le w). \tag{4.4.33}$$

From (4.4.33) it follows that

$$F_W(w) = P(-rT \le \log w)$$

or equivalently

$$F_W(w) = P\left(T \ge -\frac{1}{r}\log w\right). \tag{4.4.34}$$

From (4.4.34) it follows that

$$F_W(w) = 1 - P\left(T \le -\frac{1}{r}\log w\right). \tag{4.4.35}$$

Since

$$P(T \le t) = F_T(t)$$

and

$$F_T(t) = P_N(F_X(t))$$

then (4.4.35) implies that the distribution function $F_W(w)$ of the random variable $W = e^{-rT}$ is

$$F_W(w) = 1 - P_N\left(F_X\left(-\frac{1}{r}\log w\right)\right), \quad 0 < w < 1. \tag{4.4.36}$$

The evaluation of the characteristic function $\varphi_Y(u)$ of the stochastic model

$$Y = Le^{-rT}$$

or equivalently the stochastic model $Y = LW$ is implemented by the following way.

We have

$$\varphi_Y(u) = E\left(e^{iuLW}\right)$$

or equivalently we have

$$\varphi_Y(u) = E\big(E\big(e^{iuLW}|W\big)\big). \tag{4.4.37}$$

From (4.4.37) we get that

$$\varphi_Y(u) = \int_0^1 E\big(e^{iuLW}|W = w\big)dF_W(w). \tag{4.4.38}$$

Hence (4.4.38) implies that

$$\varphi_Y(u) = \int_0^1 E\big(e^{iuLW}|W = w\big)dF_W(w).$$

Since the independence of the random variables $L = C_1 + C_2 + \cdots + C_S$ and $T = \max(X_1, X_2, \ldots, X_N)$ implies the independence of the random variables $L = C_1 + C_2 + \cdots + C_S$ and $W = \exp[-r\max(X_1, X_2, \ldots, X_N)]$ then we get that

$$\varphi_Y(u) = \int_0^1 E\big(e^{iuwL}\big)dF_W(w). \tag{4.4.39}$$

Since the characteristic function of the random variable $L = C_1 + C_2 + \cdots + C_S$ is

$$\varphi_L(u) = P_S(\varphi_C(u)) \tag{4.4.40}$$

and (4.4.36) implies that the random variable $W = \exp[-r\max(X_1, X_2, \ldots, X_N)]$ has distribution function

$$F_W(w) = 1 - P_N\left(F_X\left(-\frac{1}{r}\log w\right)\right), \quad 0 < w < 1 \tag{4.4.41}$$

then (4.4.39), (4.4.40) and (4.4.41) imply that the characteristic function of the stochastic model

$$Y = Le^{-rT}$$

is

$$\varphi_Y(u) = \int_0^1 P_S(\varphi_C(uw))d\left[1 - P_N\left(F_X\left(-\frac{1}{r}\log\, w\right)\right)\right]. \tag{4.4.42}$$

For embedding the characteristic function (4.4.42) in the class of characteristic functions corresponding to a-unimodal distributions we make use of the following way.

We suppose that the random variable X follows the exponential distribution with distribution function

$$F_X(x) = 1 - e^{-\mu x},$$

and the random variable N follows the Sibuya distribution with probability generating function

$$P_N(z) = 1 - (1 - z)^{\gamma}, \quad 0 < \gamma \leq 1$$

and we set $a = \mu\gamma/r$, then (4.4.42) implies that the characteristic function of the stochastic model

$$Y = Le^{-rT}$$

has the form

$$\varphi_Y(u) = a \int_0^1 P_S(\varphi_C(uw))\, w^{a-1}\, dw \tag{4.4.43}$$

and the characteristic function in (4.4.43) corresponds to probability distribution which is a-unimodal.

An application of the stochastic model

$$Y = Le^{-rT},$$

where $L = C_1 + C_2 + \cdots + C_S$ and $T = \max(X_1, X_2, \ldots, X_N)$, in banking is the following.

We suppose that the random variable N denotes the number of operations of a bank which are interrupted due to the occurrence of a risk at the time point 0. The random variable X_n, $n = 1, 2, \ldots$ denotes the time required for the recovery of the nth interrupted operation of the bank. Hence the random variable $T = \max(X_1, X_2, \ldots, X_N)$ denotes the time point of recovery of the bank. Moreover we suppose that the random variable S denotes the number of loans in the portfolio of loans of the bank at the time point $T = \max(X_1, X_2, \ldots, X_N)$ and the random variable C_S denotes the size of the sth loan. Hence the random variable $L = C_1 + C_2 + \cdots + C_S$ denotes the size of the portfolio of loans of the bank at the time point $T = \max(X_1, X_2, \ldots, X_N)$.

In this case the stochastic model

$$Y = Le^{-rT}$$

denotes the present value at the time point 0 of the size $L = C_1 + C_2 + \cdots + C_S$ of the portfolio of loans of the bank at the time point $T = \max(X_1, X_2, \ldots, X_N)$ of recovery of the bank. Another application of the stochastic model

$$Y = Le^{-rT}$$

in banking arises if we use a different interpretation for the random sum $L = C_1 + C_2 + \cdots + C_S$.

We suppose that the random variable S denotes the number of investments in the portfolio of investments of the bank at the time point $T = \max(X_1, X_2, \ldots, X_N)$ of recovery of the bank and the random variable C_S denotes the market value of the sth investment of the portfolio of investments of the bank. Hence the random variable $L = C_1 + C_2 + \cdots + C_S$ denotes the market value of the portfolio of investments of the bank at the time point $T = \max(X_1, X_2, \ldots, X_N)$.

In this case the stochastic model

$$Y = Le^{-rT}$$

denotes the present value at the time point 0 of the market value $L = C_1 + C_2 + \cdots + C_S$ of the portfolio of investments of the bank corresponding at the time point $T = \max(X_1, X_2, \ldots, X_N)$ of recovery of the bank.

An application of the stochastic model

$$Y = Le^{-rT}$$

in the area of ongoing risk occurrences is the following.

We suppose that the random variable N denotes the number of ongoing occurrences of a risk at the time point 0. The random variable X_n denotes the time required for completing the duration of the nth risk occurrence which is ongoing at the time point 0. Hence the random variable $T = \max(X_1, X_2, \ldots, X_N)$ denotes the time required for completing the durations of the N risk occurrences which are ongoing at the time point 0. Moreover, we suppose that the firm facing the risk, of which N occurrences are ongoing at the time point 0, creates a portfolio of independent investments at the time point $T = \max(X_1, X_2, \ldots, X_N)$ of completing the durations of the N risk occurrences which are ongoing at the time point 0. The random variable S denotes the number of investments of the portfolio that the firm creates at the time point $T = \max(X_1, X_2, \ldots, X_N)$. The random variable C_s denotes the cost of the sth investment at the time point $T = \max(X_1, X_2, \ldots, X_N)$.

Hence the random variable $L = C_1 + C_2 + \cdots + C_S$ denotes the cost of the portfolio of investments at the time point $T = \max(X_1, X_2, \ldots, X_N)$.

In this case the stochastic model

$$Y = Le^{-rT}$$

denotes the present value at the time point 0 of the cost $L = C_1 + C_2 + \cdots + C_S$ of the portfolio of investments of the firm corresponding at the time point $T = \max(X_1, X_2, \ldots, X_N)$ of completing the durations of the N risk occurrences which are ongoing at the time point 0.

Another application of the stochastic model

$$Y = Le^{-rT}$$

in the area of ongoing risk occurrences arises if the random sum $L = C_1 + C_2 + \cdots + C_S$ is interpreted by the following way.

We suppose that the random variable S denotes the number of banks participating in the share capital of the firm at the time point $T = \max(X_1, X_2, \ldots, X_N)$ of completing the durations of the N risk occurrences which are ongoing at the time point 0. The random variable C_s denotes the size of the share capital of the firm that belongs to the sth bank at the time point $T = \max(X_1, X_2, \ldots, X_N)$.

Hence the random variable $L = C_1 + C_2 + \cdots + C_S$ denotes the size of the share capital of the firm that belongs to the S banks at the time point $T = \max(X_1, X_2, \ldots, X_N)$.

In this case the stochastic model

$$Y = Le^{-rT}$$

denotes the present value at the time point 0 of the size $L = C_1 + C_2 + \cdots + C_S$ of the share capital of the firm that belongs to the S banks at the time point $T = \max(X_1, X_2, \ldots, X_N)$ of completing the durations of the N risk occurrences which are ongoing at the point 0.

An interesting modification of the stochastic model

$$Y = Le^{-rT},$$

where $L = C_1 + C_2 + \cdots + C_S$ and $T = \max(X_1, X_2, \ldots, X_N)$ is implemented by the following way.

Let N be a discrete random variable with values in the set $\mathbf{N} = \{1, 2, \ldots\}$ and probability generating function $P_N(z)$.

We suppose that $\{X_n, n = 1, 2, \ldots\}$ is a sequence of continuous, positive, independent, and identically distributed random variables. The random variables of the sequence are equally distributed with the random variable X having distribution function $F_X(x)$ and we set $T = \max(X_1, X_2, \ldots, X_N)$.

Let Π be a continuous and positive random variable with characteristic function $\varphi_\Pi(u)$, and U be a continuous and positive random variable with characteristic function $\varphi_U(u)$.

We consider the positive real number r and we set

$$V = U + \Pi \exp[-r \max(X_1, X_2, \ldots, X_N)]$$

or equivalently

$$V = U + \Pi e^{-rT}$$

The establishment of properties and applications in risk management of the stochastic model

$$V = U + \Pi e^{-rT}$$

is of particular practical and theoretical interest. An interpretation in the area of continuous discounting of that model is the following.

We suppose that each random variable of the sequence $\{X_n, n = 1, 2, \ldots\}$ denotes time, the random variable Π denotes a cash flow, the random variable U denotes a cash flow and the positive real number r denotes force of interest. In this case the stochastic model

$$Y = U + \Pi e^{-rT}$$

denotes the sum of the cash flow U corresponding to the time point 0 and the present value at the time point 0 of the cash flow Π corresponding to the time point $T = \max(X_1, X_2, \ldots, X_N)$.

The following result establishes sufficient conditions for evaluating the characteristic function $\varphi_V(u)$ of the stochastic model

$$V = U + \Pi e^{-rT}.$$

□

Theorem 4.4.2 *Let N be a discrete random variable with values in the set $\mathbf{N} = \{1, 2, \ldots\}$ and probability generating function $P_N(z)$.*

We suppose that $\{X_n, n = 1, 2, \ldots\}$ is a sequence of continuous, positive, independent, and identically distributed random variables. The sequence of random variables are equally distributed with the random variable X having distribution function $F_X(x)$ and we set $T = \max(X_1, X_2, \ldots, X_N)$.

Let Π be a continuous and positive random variable with characteristic function $\varphi_\Pi(u)$ and U be a continuous and positive random variable with characteristic function $\varphi_U(u)$.

We consider the positive real number r and we set

$$V = U + \Pi e^{-rT}.$$

If N, $\{X_n, n = 1, 2, \ldots\}$, Π *and* U *are independent then the characteristic function of the stochastic model* $V = U + \Pi e^{-rT}$ *is*

$$\varphi_V(u) = \varphi_U(u) \int_0^1 \varphi_\Pi(uw) d\left[1 - P_N\left(F_X\left(-\frac{1}{r}\log w\right)\right)\right].$$

Proof The independence of N, $\{X_n, n = 1, 2, \ldots\}$, Π and U implies the independence of the random variable N and the sequence of the random variables $\{X_n,\ n = 1, 2, \ldots\}$.

Hence the random variable $T = \max(X_1, X_2, \ldots, X_N)$ has distribution function

$$F_T(t) = P_N(F_X(t))$$

and the random variable $W = e^{-rT}$ has distribution function

$$F_W(w) = 1 - P_N\left(F_X\left(-\frac{1}{r}\log w\right)\right), \quad 0 < w < 1.$$

Let $F_U(\upsilon)$ be the distribution function of the random variable U, $F_\Pi(\pi)$ be the distribution function of the random variable Π,

$$F_T(t) = P_N(F_X(t))$$

be the distribution function of the random variable $T = \max(X_1, X_2, \ldots, X_N)$ and $F_{U,\Pi,T}(\upsilon, \pi, t)$ be the distribution function of the vector (U, Π, T).

The establishment of the relationship

$$F_{U,\Pi,T}(\upsilon, \pi, t) = F_U(\upsilon) F_\Pi(\pi) F_T(t)$$

implies the independence of the random variables U, Π, T.

We have

$$F_{U,\Pi,T}(\upsilon, \pi, t) = P(U \le \upsilon,\ \Pi \le \pi,\ T \le t)$$

or equivalently

$$F_{U,\Pi,T}(\upsilon, \pi, t) = \sum_{n=1}^{\infty} P(U \le \upsilon,\ \Pi \le \pi,\ T \le t | N = n) P(N = n). \tag{4.4.44}$$

From (4.4.44) it follows that

$$F_{U,\Pi,T}(\upsilon, \pi, t) = \sum_{n=1}^{\infty} P(U \leq \upsilon,\ \Pi \leq \pi,\ \max(X_1, \ldots, X_N) \leq t | N = n) P(N = n). \tag{4.4.45}$$

Hence (4.4.45) implies that

$$F_{U,\Pi,T}(\upsilon, \pi, t) = \sum_{n=1}^{\infty} P(U \leq \upsilon,\ \Pi \leq \pi,\ \max(X_1, \ldots, X_n) \leq t | N = n) P(N = n). \tag{4.4.46}$$

From (4.4.46) it follows that

$$F_{U,\Pi,T}(\upsilon, \pi, t) = \sum_{n=1}^{\infty} P(U \leq \upsilon,\ \Pi \leq \pi,\ X_1 \leq t, \ldots, X_n \leq t | N = n) P(N = n) \tag{4.4.47}$$

Since the independence of N, $\{X_n, n = 1, 2, \ldots\}$, Π and U implies the independence of the random variables N, $X_1, \ldots, X_n, \Pi, U$ then (4.4.47) has the form

$$F_{U,\Pi,T}(\upsilon, \pi, t) = \sum_{n=1}^{\infty} P(U \leq \upsilon,\ \Pi \leq \pi,\ X_1 \leq t, \ldots, X_n \leq t) P(N = n). \tag{4.4.48}$$

Since the independence of the random variables $N, X_1, \ldots, X_n, \Pi, U$ implies the independence of the random variables $X_1, \ldots, X_n, \Pi, U$ then (4.4.48) has the form

$$F_{U,\Pi,T}(\upsilon, \pi, t) = \sum_{n=1}^{\infty} P(U \leq \upsilon) P(\Pi \leq \pi) P(X_1 \leq t) \ldots P(X_n \leq t) P(N = n)$$

or equivalently the form

$$F_{U,\Pi,T}(\upsilon, \pi, t) = F_U(\upsilon) F_{\Pi}(\pi) \sum_{n=1}^{\infty} F_X^n(t) P(N = n). \tag{4.4.49}$$

From (4.4.49) we get that

$$F_{U,\Pi,T}(\upsilon, \pi, t) = F_U(\upsilon) F_{\Pi}(\pi) P_N(F_X(t))$$

or equivalently we get that

$$F_{U,\Pi,T}(\upsilon, \pi, t) = F_U(\upsilon) F_{\Pi}(\pi) F_T(t). \tag{4.4.50}$$

Hence (4.4.50) implies that the random variables U, Π, T are independent. The independence of these random variables implies the independence of the random variables U, Π, $W = \mathrm{e}^{-rT}$.

Moreover, the independence of the random variables U, Π, $W = \mathrm{e}^{-rT}$ implies the independence of the random variables Π, W.

If $\varphi_{\Pi W}(u)$ is the characteristic function of the random variable ΠW then we get that

$$\varphi_{\Pi W}(u) = E\left(e^{iu\Pi W}\right)$$

or equivalently

$$\varphi_{\Pi W}(u) = E\left(E\left(e^{iu\Pi W}|W\right)\right). \tag{4.4.51}$$

From (4.4.51) we get that

$$\varphi_{\Pi W}(u) = \int_0^1 E\left(e^{iu\Pi W}|W = w\right) dF_W(w)$$

or equivalently

$$\varphi_{\Pi W}(u) = \int_0^1 E\left(e^{iu\Pi W}|W = w\right) d\left[1 - P_N\left(F_X\left(-\frac{1}{r}\log w\right)\right)\right]. \tag{4.4.52}$$

Since the random variables Π, W are independent then (4.4.52) has the form

$$\varphi_{\Pi W}(u) = \int_0^1 \varphi_{\Pi}(uw) d\left[1 - P_N\left(F_X\left(-\frac{1}{r}\log w\right)\right)\right].$$

If $\varphi_V(u)$ is the characteristic function of the stochastic model

$$V = U + \Pi \mathrm{e}^{-rT}$$

or equivalently the stochastic model

$$V = U + \Pi W$$

then we get that

$$\varphi_V(u) = E\left(e^{iu(U+\Pi W)}\right). \tag{4.4.53}$$

Hence (4.4.53) implies that

$$\varphi_V(u) = E\Big(E\Big(e^{iu(U+\Pi W)}|W\Big)\Big). \tag{4.4.54}$$

From (4.4.54) it follows that

$$\varphi_V(u) = \int_0^1 E\Big(e^{iu(U+\Pi W)}|W = w\Big)dF_W(w). \tag{4.4.55}$$

and (4.4.55) implies that

$$\varphi_V(u) = \int_0^1 E\Big(e^{iu(U+\Pi W)}|W = w\Big)d\left[1 - P_N\left(F_X\left(-\frac{1}{r}\log w\right)\right)\right]. \tag{4.4.56}$$

Hence (4.4.56) implies that

$$\varphi_V(u) = \int_0^1 E(e^{iuU+iuw\Pi}|W = w)d\left[1 - P_N\left(F_X\left(-\frac{1}{r}\log w\right)\right)\right]. \tag{4.4.57}$$

Since the random variables U, Π, $W = \mathrm{e}^{-rT}$ are independent then (4.4.57) has the form

$$\varphi_V(u) = \int_0^1 E(e^{iuU+iuw\Pi})d\left[1 - P_N\left(F_X\left(-\frac{1}{r}\log w\right)\right)\right]. \tag{4.4.58}$$

Since the independence of the random variables U, Π, $W = \mathrm{e}^{-rT}$ implies the independence of the random variables U, Π then (4.4.58) has the form

$$\varphi_V(u) = \int_0^1 E(e^{iuU})E(e^{iuw\Pi})d\left[1 - P_N\left(F_X\left(-\frac{1}{r}\log w\right)\right)\right]. \tag{4.4.59}$$

From (4.4.59) it follows that

$$\varphi_V(u) = E(e^{iuU})\int_0^1 E(e^{iuw\Pi})d\left[1 - P_N\left(F_X\left(-\frac{1}{r}\log w\right)\right)\right]. \tag{4.4.60}$$

Since

$$\varphi_U(u) = E(e^{iuU})$$

is the characteristic function of the random variable U and

$$\varphi_\Pi(u) = E(e^{iu\Pi})$$

is the characteristic function of the random variable Π then (4.4.60) implies that the characteristic function of the stochastic model

$$V = U + \Pi e^{-rT},$$

where $T = \max(X_1, X_2, \ldots, X_N)$, is

$$\varphi_V(u) = \varphi_U(u) \int_0^1 \varphi_\Pi(uw) d\left[1 - P_N\left(F_X\left(-\frac{1}{r}\log w\right)\right)\right]$$

or equivalently

$$\varphi_V(u) = \varphi_U(u)\varphi_{\Pi W}(u).$$

An application of the stochastic model

$$V = U + \Pi e^{-rT},$$

where $T = \max(X_1, X_2, \ldots, X_N)$, in risk management is the following.

We suppose that the random variable N denotes the number of ongoing occurrences of a risk at the time moment 0. The random variable X_n denotes the time required for completing the duration of the nth risk occurrence which is ongoing at the time point 0. Hence the random variable $T = \max(X_1, X_2, \ldots, X_N)$ denotes the time required for completing the durations of the N occurrences of the risk which are ongoing at the time point 0. Moreover we suppose that the firm facing a risk, of which N occurrences are ongoing at the time point 0, liquidates an investment at the time point $T = \max(X_1, X_2, \ldots, X_N)$ of completing the durations of the N occurrences of the risk which are ongoing at the time point 0. The random variable Π denotes the price of liquidation of the investment at the time point $T = \max(X_1, X_2, \ldots, X_N)$.

The random variable Πe^{-rT} denotes the present value at the time point 0 of the price of liquidation Π of the investment at the time point $T = \max(X_1, X_2, \ldots, X_N)$.

If the random variable U denotes the cash reserve of the firm at the time point 0 then the stochastic model

$$V = U + \Pi e^{-rT}$$

denotes the total cash reserve of the firm at the time point 0.

Another application of the stochastic model

$$V = U + \Pi e^{-rT}$$

in the area of ongoing risk occurrences is the following. We suppose that the random variable U denotes the size of a loan that the firm undertakes at the time point 0 and the random variable Π denotes the size of another loan that the firm undertakes at the time point $T = \max(X_1, X_2, \ldots, X_N)$.

The random variable Πe^{-rT} denotes the present value at the time point 0 of the size Π of the loan that the firm undertakes at the time point $T = \max(X_1, X_2, \ldots, X_N)$.

Hence the stochastic model

$$V = U + \Pi e^{-rT}$$

denotes the total loan obligation of the firm at the time point 0.

The following result concentrates on the investigation of the characteristic function $\varphi_V(u)$ of a particular case of the stochastic model

$$V = U + \Pi e^{-rT}.$$

□

Theorem 4.4.3 *Let N be a discrete random variable with values in the set $\mathbf{N} = \{1, 2, \ldots\}$ following the Sibuya distribution with probability generating function*

$$P_N(z) = 1 - (1 - z)^{\gamma}, \quad 0 < \gamma \leq 1.$$

We suppose that $\{X_n, n = 1, 2, \ldots\}$ is a sequence of continuous, positive, independent, and identically distributed random variables. The random variables of the sequence are equally distributed with the random variable X following the exponential distribution with distribution function

$$F_X(x) = 1 - e^{-\mu x}, \quad x > 0 \quad \mu > 0$$

and we set $T = \max(X_1, X_2, \ldots, X_N)$.

Let Π be a continuous and positive random variable with characteristic function $\varphi_\Pi(u)$ and finite mean. Let U be a continuous and positive random variable with characteristic function $\varphi_U(u)$ and finite mean. We consider the stochastic model

$$V = U + \Pi e^{-rT},$$

where $r > 0$.

We suppose that N, $\{X_n, n = 1, 2, \ldots\}$, Π and U *are independent. The characteristic function of the stochastic model*

$$V = U + \Pi e^{-rT}$$

is

$$\varphi_V(u) = \varphi_U(u) \exp\left(a \int\limits_0^u \frac{\varphi_U(y) - 1}{y} dy \right),$$

where $a = \mu\gamma/r$, *if, and, only, if the random variables* V, Π *are equally distributed.*

Proof From Theorem 4.4.2 it follows that the characteristic function of the stochastic model

$$V = U + \Pi e^{-rT}$$

is

$$\varphi_V(u) = \varphi_U(u) \int\limits_0^1 \varphi_\Pi(uw) d\left[1 - P_N\left(F_X\left(-\frac{1}{r}\log w\right)\right)\right]. \quad (4.4.61)$$

Since the probability generating function of the random variable N is

$$P_N(z) = 1 - (1 - z)^\gamma$$

and the distribution function of the random variable X is

$$F_X(x) = 1 - e^{-\mu x}, \quad x > 0 \quad \mu > 0$$

then the distribution function of the random variable $T = \max(X_1, X_2, \ldots, X_N)$ is

$$F_T(t) = 1 - e^{-\mu\gamma t}, \quad t > 0 \quad \mu\gamma > 0$$

and the distribution function of the random variable $W = e^{-rT}$ is

$$F_W(w) = w^a, \quad 0 < w < 1,$$

where $a = \mu\gamma/r$.

Hence (4.4.61) has the form

$$\varphi_V(u) = \varphi_U(u) a \int_0^1 \varphi_\Pi(uw) w^{a-1} dw$$

or equivalently the form

$$\varphi_V(u) = \varphi_U(u) \frac{a}{u^a} \int_0^u \varphi_\Pi(w) w^{a-1} dw. \tag{4.4.62}$$

If the random variables V, Π are equally distributed ναι then we get that

$$\varphi_V(u) = \varphi_\Pi(u). \tag{4.4.63}$$

From (4.4.62) and (4.4.63) we get that

$$\varphi_V(u) = \varphi_U(u) \cdot \frac{a}{u^a} \int_0^u \varphi_V(w) w^{a-1} dw. \tag{4.4.64}$$

If we multiply both sides of the integral equation (4.4.64) by

$$\frac{u^a}{\varphi_U(u)}, \quad \varphi_U(u) \neq 0$$

then we get the integral equation

$$u^a \frac{\varphi_V(u)}{\varphi_U(u)} = a \int_0^u \varphi_V(w) w^{a-1} dw. \tag{4.4.65}$$

If we differentiate the integral equation (4.4.65) then we get the differential equation

$$au^{a-1} \frac{\varphi_V(u)}{\varphi_U(u)} + \frac{u^a}{\varphi_U(u)} \cdot \frac{d\varphi_V(u)}{du} - u^a \frac{\varphi_V(u)}{\varphi_U^2(u)} \cdot \frac{d\varphi_U(u)}{du} = a\varphi_V(u) u^{a-1}. \tag{4.4.66}$$

If $u \neq 0$ then the differential equation (4.4.66) has the form

$$\frac{\varphi_V(u)}{\varphi_U(u)} + \frac{u}{a\varphi_U(u)} \cdot \frac{d\varphi_V(u)}{du} - \frac{u\varphi_V(u)}{a\varphi_U^2(u)} \cdot \frac{d\varphi_U(u)}{du} = \varphi_V(u). \tag{4.4.67}$$

Moreover, the differential equation (4.4.67) can be written in the form

$$\frac{1}{\varphi_V(u)}\cdot\frac{d\varphi_V(u)}{du}-\frac{1}{\varphi_U(u)}\cdot\frac{d\varphi_U(u)}{du}=a\frac{\varphi_U(u)-1}{u} \tag{4.4.68}$$

for u such that $\varphi_V(u)\neq 0$.

If we integrate both sides of (4.4.68) then we get that

$$\int_0^u\frac{d\varphi_V(y)}{\varphi_V(y)}-\int_0^u\frac{d\varphi_U(y)}{\varphi_U(y)}=a\int_0^u\frac{\varphi_U(y)-1}{y}dy. \tag{4.4.69}$$

From (4.4.69) we get that

$$\log\varphi_V(u)-\log\varphi_U(u)=a\int_0^u\frac{\varphi_U(y)-1}{y}dy$$

or equivalently

$$\log\frac{\varphi_V(u)}{\varphi_U(u)}=a\int_0^u\frac{\varphi_U(y)-1}{y}dy \tag{4.4.70}$$

with u such that $\varphi_U(u)\neq 0$.

From (4.4.70) we get that

$$\log\frac{\varphi_V(u)}{\varphi_U(u)}=\log\exp\left(a\int_0^u\frac{\varphi_U(y)-1}{y}dy\right)$$

or equivalently

$$\frac{\varphi_V(u)}{\varphi_U(u)}=\exp\left(a\int_0^u\frac{\varphi_U(y)-1}{y}dy\right)$$

Hence

$$\varphi_V(u)=\varphi_U(u)\exp\left(a\int_0^u\frac{\varphi_U(y)-1}{y}dy\right).$$

The inverse is obvious. □

4.5 Recovery Time of a Partially Damaged System and Present Value of a Continuous Uniform Cash Flow

Let $\{C_s, s = 1, 2, \ldots\}$ be a sequence of continuous, positive, independent and identically distributed random variables. We consider the discrete random variable S with values in the set $\mathbf{N}_0 = \{0, 1, 2, \ldots\}$ and probability generating function $P_S(z)$.

We suppose that the random variable S is independent of the sequence of continuous, positive, independent, and identically distributed random variables $\{C_s, s = 1, 2, \ldots\}$ and set $L = C_1 + C_2 + \cdots + C_S$.

Let $\{X_n, n = 1, 2, \ldots\}$ be a sequence of continuous, positive, independent, and identically distributed random variables. We consider the discrete random variable N with values in the set $\mathbf{N} = \{1, 2, \ldots\}$ and probability generating function $P_N(z)$.

We suppose that the random variable N is independent of the sequence of continuous, positive, independent, and identically distributed random variables $\{X_n, n = 1, 2, \ldots\}$ and we set $T = \max(X_1, X_2, \ldots, X_N)$.

We consider the stochastic model

$$Y = L\frac{1 - \mathrm{e}^{-rT}}{r},$$

where r is a positive real number. The purpose of the present section is the investigation and applications in risk management of the above stochastic model. An interpretation in the area of continuous discounting of that model is the following.

We suppose that each random variable of the sequence $\{X_n, n = 1, 2, \ldots\}$ denotes time, each random variable of the sequence $\{C_s, s = 1, 2, \ldots\}$ denotes a cash flow and the positive real number r denotes force of interest. In this case the stochastic model

$$Y = L\frac{1 - \mathrm{e}^{-rT}}{r}$$

denotes the present value at the time point 0 of the continuous uniform cash flow with rate of payment $L = C_1 + C_2 + \cdots + C_S$ and duration $T = \max(X_1, X_2, \ldots, X_N)$.

The following result establishes sufficient conditions for evaluating the characteristic function of the model

$$Y = L\frac{1 - \mathrm{e}^{-rT}}{r}.$$

Theorem 4.5.1 *Let* $\{C_s, s = 1, 2, \ldots\}$ *be a sequence of continuous, positive, independent, and identically distributed random variables. The random variables of the sequence* $\{C_s, s = 1, 2, \ldots\}$ *are equally distributed with the random variable* C *having characteristic function* $\varphi_C(u)$.

We consider the discrete random variable S *with values in the set* $\mathbf{N}_0 = \{0, 1, 2, \ldots\}$, *probability generating function* $P_S(z)$ *and we set* $L = C_1 + C_2 + \cdots + C_S$.

Let $\{X_n, n = 1, 2, \ldots\}$ *be a sequence of continuous, positive, independent, and identically distributed random variables. The random variables of the sequence* $\{X_n, n = 1, 2, \ldots\}$ *are equally distributed with the random variable* X *having distribution function* $F_X(x)$.

We consider the discrete random variable N *with values in the set* $\mathbf{N} = \{1, 2, \ldots\}$, *probability generating function* $P_N(z)$ *and we set* $T = \max(X_1, X_2, \ldots, X_N)$.

We consider the stochastic model

$$Y = L\frac{1 - e^{-rT}}{r}$$

where r *is positive real number. If* $\{C_s, s = 1, 2, \ldots\}$, S, $\{X_n, n = 1, 2, \ldots\}$ *and* N *are independent then the characteristic function of the stochastic model*

$$Y = L\frac{1 - e^{-rT}}{r}$$

is

$$\varphi_Y(u) = \int_0^{1/r} P_S(\varphi_C(uw))dP_N\left(F_X\left(-\frac{1}{r}\log(1 - rw)\right)\right)$$

or equivalently

$$\varphi_Y(u) = \int_0^{1} P_S(\varphi_C(uw/r))dP_N\left(F_X\left(-\frac{1}{r}\log(1 - w)\right)\right).$$

Proof We consider the random variable $L = C_1 + C_2 + \cdots + C_S$ and the random variable $T = \max(X_1, X_2, \ldots, X_N)$.

From Theorem 4.4.1 it follows that the random variables $L = C_1 + C_2 + \cdots + C_S$, $T = \max(X_1, X_2, \ldots, X_N)$ are independent,

$$\varphi_L(u) = P_S(\varphi_C(u))$$

is the characteristic function of the random variable $L = C_1 + C_2 + \cdots + C_S$ and

$$F_T(t) = P_N(F_X(t))$$

is the distribution function of the random variable $T = \max(X_1, X_2, \ldots, X_N)$.

The independence of the random variables $L = C_1 + C_2 + \cdots + C_S$ and $T = \max(X_1, X_2, \ldots, X_N)$ implies the independence of the random variables $L = C_1 + C_2 + \cdots + C_S$, $W = \{1 - \exp[-r\max(X_1, X_2, \ldots, X_N)]\}/r$.

The evaluation of the characteristic function of the stochastic model

$$Y = L\frac{1 - \mathrm{e}^{-rT}}{r}$$

or equivalently the stochastic model $Y = LW$ requires the evaluation of the distribution function $F_W(w)$ of the random variable

$$W = \frac{1 - \mathrm{e}^{-rT}}{r}.$$

We have

$$F_W(w) = P(W \le w)$$

or equivalently we have

$$F_W(w) = P\left(\frac{1 - e^{-rT}}{r} \le w\right). \tag{4.5.1}$$

From (4.5.1) it follows that

$$F_W(w) = P(1 - \mathrm{e}^{-rT} \le rw)$$

or equivalently it follows that

$$F_W(w) = P(-\mathrm{e}^{-rT} \le rw - 1). \tag{4.5.2}$$

Hence (4.5.2) implies that

$$F_W(w) = P(\mathrm{e}^{-rT} \ge 1 - rw)$$

or equivalently

$$F_W(w) = P(-rT \geq \log(1 - rw)). \tag{4.5.3}$$

From (4.5.3) it follows that

$$F_W(w) = P\left(T \leq -\frac{1}{r}\log(1 - rw)\right). \tag{4.5.4}$$

Since

$$P(T \leq t) = P_N(F_X(t))$$

then (4.5.4) implies that the distribution function $F_W(w)$ of the random variable

$$W = \frac{1 - e^{-rT}}{r}$$

is

$$F_W(w) = P_N\left(F_X\left(-\frac{1}{r}\log(1 - rw)\right)\right).$$

The independence of the random variables L, W and the distribution function $F_W(w)$ permit the evaluation of the characteristic function $\varphi_Y(u)$ of the stochastic model

$$Y = L\frac{1 - e^{-rT}}{r}$$

in the following way.

We have

$$\varphi_Y(u) = E\left(e^{iuLW}\right)$$

or equivalently we have

$$\varphi_Y(u) = E\left(E\left(e^{iuLW}|W\right)\right). \tag{4.5.5}$$

From (4.5.5) it follows that

$$\varphi_Y(u) = \int_0^{1/r} E\left(e^{iuLW}|W = w\right)dF_W(w)$$

or equivalently

$$\varphi_Y(u) = \int_0^{1/r} E\left(e^{iuwL} | W = w\right) dF_W(w). \tag{4.5.6}$$

The independence of the random variables L, W implies that (4.5.6) has the form

$$\varphi_Y(u) = \int_0^{1/r} E\left(e^{iuwL}\right) dF_W(w). \tag{4.5.7}$$

Since

$$\varphi_L(u) = E\left(e^{iuL}\right)$$

then (4.5.7) has the form

$$\varphi_Y(u) = \int_0^{1/r} \varphi_L(uw) dF_W(w). \tag{4.5.8}$$

Since

$$\varphi_L(u) = P_S(\varphi_C(u))$$

and

$$F_W(w) = P_N\left(F_X\left(-\frac{1}{r}\log(1 - rw)\right)\right)$$

then (4.5.8) has the form

$$\varphi_Y(u) = \int_0^{1/r} P_S(\varphi_C(uw)) dP_N\left(F_X\left(-\frac{1}{r}\log(1 - rw)\right)\right). \tag{4.5.9}$$

From (4.5.9) it follows that

$$\varphi_Y(u) = \int_0^{1} P_S\left(\varphi_C\left(^{uw}/_{r}\right)\right) dP_N\left(F_X\left(-\frac{1}{r}\log(1 - w)\right)\right). \tag{4.5.10}$$

For embedding the characteristic function (4.5.10) in the class of characteristic functions corresponding to v-unimodal probability distributions we work as follows.

We suppose that the random variable X follows the exponential distribution with distribution function

$$F_X(x) = 1 - e^{-\mu x},$$

the random variable N follows the Sibuya distribution with probability generating function

$$P_N(z) = 1 - (1 - z)^{\gamma}$$

and we set $v = \mu\gamma/r$ then (4.5.10) implies that the characteristic function of the stochastic model

$$Y = L\frac{1 - \mathrm{e}^{-rT}}{r}$$

is

$$\varphi_Y(u) = v\int_0^1 P_S\left(\varphi_C\left({}^{uw}/_{r}\right)\right)(1 - w)^{v-1}dw$$

and that characteristic function corresponds to a v-unimodal probability distribution.

An application of the stochastic model

$$Y = L\frac{1 - \mathrm{e}^{-rT}}{r}$$

where $L = C_1 + C_2 + \cdots + C_S$ and $T = \max(X_1, X_2, \ldots, X_N)$ in industrial activities is the following.

We suppose that the random variable N denotes the number of production lines of an industrial firm which are interrupted due to the occurrence of a risk at the time 0. The random variable X_n denotes the time required for recovery of the nth interrupted production line of the industrial firm. Hence the random variable $T = \max(X_1, X_2, \ldots, X_N)$ denotes the recovery time point of the industrial firm. Moreover, we suppose that the random variable S denotes the number of different lost incomes of the industrial firm at every time point of the interval $[0, T]$ with $T = \max(X_1, X_2, \ldots, X_N)$, due to the partial production activity of the industrial firm in the time interval $[0, T]$.

The random variable C_s denotes the size of the sth lost income at every time point of the interval $[0, T]$.

Hence the random variable $L = C_1 + C_2 + \cdots + C_S$ denotes the total size of different lost incomes of the industrial firm at every time point of the interval $[0, T]$.

In this case the stochastic model

$$Y = L\frac{1 - e^{-rT}}{r}$$

denotes the present value at the time point 0 of the total size of different lost incomes of the industrial firm in the time interval $[0, T]$.

Another application of the stochastic model

$$Y = L\frac{1 - e^{-rT}}{r}$$

in the area of industrial activities arises if the random sum $L = C_1 + C_2 + \cdots + C_S$ is interpreted by the following way. We suppose that the random variable S denotes the number of different expenses of the firm at any time point of the interval $[0, T]$, with $T = \max(X_1, X_2, \ldots, X_N)$, which are implemented for the recovery of the production activity. The random variable C_s denotes the size of the sth expense corresponding at any time point of the interval $[0, T]$.

Hence the random variable $L = C_1 + C_2 + \cdots + C_S$ denotes the total size of different expenses of the firm at any time point of the interval $[0, T]$.

In this case the stochastic model

$$Y = L\frac{1 - e^{-rT}}{r}$$

denotes the present value at the time point 0 of the total size of different expenses of the firm in the time interval $[0, T]$.

An application of the stochastic model

$$Y = L\frac{1 - e^{-rT}}{r}$$

in the area of ongoing risk occurrences is the following. We suppose that the random variable N denotes the number of ongoing occurrences of a risk at the time point 0. The random variable X_n denotes the time required for completing of the nth occurrence of the risk which is ongoing at the time point 0. Hence the random variable $T = \max(X_1, X_2, \ldots, X_N)$ denotes the time required for completing the durations of the N occurrences of the risk which are ongoing at the time point 0. Moreover, we suppose that the random variable S denotes the number of lost incomes of the firm at any time point of the interval $[0, T]$, with $T = \max(X_1, X_2, \ldots, X_N)$, due to the N risk occurrences which are ongoing at the time point 0 and which risk occurrences are completed in the time interval $[0, T]$.

The random variable C_s denotes the size of the sth lost income at any time point of the interval $[0, T]$.

Hence the random variable $L = C_1 + C_2 + \cdots + C_S$ denotes the total size of different lost incomes of the firm at any time point of the interval $[0, T]$.

In this case the stochastic model

$$Y = L\frac{1 - e^{-rT}}{r}$$

denotes the present value at the time point 0 of the total size of different lost incomes of the firm in the time interval $[0, T]$.

An application of the stochastic model

$$Y = L\frac{1 - e^{-rT}}{r}$$

in the area of ongoing risk occurrences arises if the random sum $L = C_1 + C_2 + \cdots + C_S$ is interpreted in the following way. We suppose that that the random variable S denotes the number of different expenses of the firm at any time point of the interval $[0, T]$, with $T = \max(X_1, X_2, \ldots, X_N)$, for treating of the N risk occurrences which are ongoing at the time point 0. The random variable C_s denotes the size of the sth expense which is realized at any time point of the interval $[0, T]$.

Hence the random variable $L = C_1 + C_2 + \cdots + C_S$ denotes the total size of different expenses of the firm at any time point of the interval $[0, T]$.

In this case the stochastic model

$$Y = L\frac{1 - e^{-rT}}{r}$$

denotes the present value at the time point 0 of the total size of different expenses in the time interval $[0, T]$. □

4.6 Time of First Damage for a System Threatened by a Random Number of Risks and Present Value of a Single Cash Flow

Let $\{C_s, s = 1, 2, \ldots\}$ be a sequence of continuous, positive, independent, and identically distributed random variables. We consider the discrete random variable S with values in the set $\mathbf{N}_0 = \{0, 1, 2, \ldots\}$ and probability generating function $P_S(z)$. we suppose that the random variable S is independent of the sequence of continuous, positive, independent, and identically distributed random variables $\{C_s, s = 1, 2, \ldots\}$ and we set $L = C_1 + C_2 + \cdots + C_S$.

Let $\{X_n, n = 1, 2, \ldots\}$ be a sequence of continuous, positive, independent, and identically distributed random variables. We consider the discrete random variable N with values in the set $\mathbf{N} = \{1, 2, \ldots\}$ and probability generating function $P_N(z)$.

We suppose that the random variable N is independent of the sequence of continuous, positive, independent, and identically distributed random variables $\{X_n, n = 1, 2, \ldots\}$ and we set $T = \min(X_1, X_2, \ldots, X_N)$.

We consider the stochastic model

$$Y = Le^{-rT}$$

where r is positive real number. The purpose of the present section is the investigation and applications in risk management of the above stochastic model. An interpretation in the area of continuous discounting of that model is the following.

We suppose that each random variable of the sequence $\{X_n, n = 1, 2, \ldots\}$ denotes time and each random variable of the sequence $\{C_s, s = 1, 2, \ldots\}$ denotes a cash flow and the positive real number r denotes force of interest. In this case the stochastic model

$$Y = Le^{-rT}$$

denotes the present value at the time point 0 of the cash flow $L = C_1 + C_2 + \cdots + C_S$ corresponding at the time point $T = \min(X_1, X_2, \ldots, X_N)$.

The following result establishes sufficient conditions for evaluating the characteristic function of the stochastic model

$$Y = Le^{-rT}.$$

Theorem 4.6.1 *Let $\{C_s, s = 1, 2, \ldots\}$ be a sequence of continuous, positive, independent, and identically distributed random variables The random variables of the sequence $\{C_s, s = 1, 2, \ldots\}$ are equally distributed with the random variable C having characteristic function*

$$\varphi_C(u). \tag{4.6.1}$$

We consider the discrete random variable S with values in the set $\mathbf{N}_0 = \{0, 1, 2, \ldots\}$ and probability generating function

$$P_S(z) \tag{4.6.2}$$

and we set $L = C_1 + C_2 + \cdots + C_S$.

Let $\{X_n, n = 1, 2, \ldots\}$ it be a sequence of continuous, positive, independent, and identically distributed random variables. The random variables of the sequence $\{X_n, n = 1, 2, \ldots\}$ are equally distributed with the random variable X having distribution function

$$F_X(x). \tag{4.6.3}$$

We consider the discrete random variable N *with values in the set* $\mathbf{N} = \{1, 2, \ldots\}$ *and probability generating function*

$$P_N(z) \tag{4.6.4}$$

and we set $T = \min(X_1, X_2, \ldots, X_N)$.

We consider the stochastic model

$$Y = Le^{-rT}$$

where r *is a positive real number. If* $\{C_s, s = 1, 2, \ldots\}$, S, $\{X_n, n = 1, 2, \ldots\}$ *and* N *are independent then the characteristic function of the stochastic model*

$$Y = Le^{-rT}$$

is

$$\varphi_Y(u) = \int_0^1 P_S(\varphi_C(uw)) dP_N\left(1 - F_X\left(-\frac{1}{r}\log w\right)\right).$$

Proof We consider the random variable $L = C_1 + C_2 + \cdots + C_S$.

The independence of $\{C_s, s = 1, 2, \ldots\}$, S, $\{X_n, n = 1, 2, \ldots\}$ and N implies the independence of S and $\{C_s, s = 1, 2, \ldots\}$.

Hence (2.5.6), (4.6.1) and (4.6.2) it follows that the characteristic function of the random variable $L = C_1 + C_2 + \cdots + C_S$ is

$$\varphi_L(u) = P_S(\varphi_C(u)).$$

Let $\varphi_T(\xi)$ be the characteristic function of the random variable $T = \min(X_1, X_2, \ldots, X_N)$ and $\varphi_{L,T}(u, \xi)$ is the characteristic function of the vector (L, T) of the random variables $L = C_1 + C_2 + \cdots + C_S$ and $T = \min(X_1, X_2, \ldots, X_N)$.

The independence of the random variables $L = C_1 + C_2 + \cdots + C_S$ and $T = \min(X_1, X_2, \ldots, X_N)$ is required for evaluating the characteristic function $\varphi_Y(u)$ of the stochastic model

$$Y = Le^{-rT}.$$

The establishment of the relationship

$$\varphi_{L,T}(u, \xi) = \varphi_L(u)\varphi_T(\xi) \tag{4.6.5}$$

implies the independence of the random variables $L = C_1 + C_2 + \cdots + C_S$ and $T = \min(X_1, X_2, \ldots, X_N)$.

We have

$$\varphi_{L,T}(u, \xi) = E\left(e^{iuL+i\xi T}\right)$$

or equivalently

$$\varphi_{L,T}(u, \xi) = E\left(E\left(e^{iuL+i\xi T}|S\right)\right). \tag{4.6.6}$$

From (4.6.6) it follows that

$$\varphi_{L,T}(u, \xi) = \sum_{s=0}^{\infty} E\left(e^{iuL+i\xi T}|S = s\right)P(S = s). \tag{4.6.7}$$

Hence (4.6.7) implies that

$$\varphi_{L,T}(u, \xi) = \sum_{s=0}^{\infty} E\left(e^{iu(C_1+C_2+\cdots+C_s)+i\xi T}|S = s\right)P(S = s)$$

or equivalently

$$\varphi_{L,T}(u, \xi) = \sum_{s=0}^{\infty} E\left(e^{iuC_1+\cdots+iuC_s+i\xi T}|S = s\right)P(S = s). \tag{4.6.8}$$

From (4.6.8) implies that

$$\varphi_{L,T}(u, \xi) = \sum_{s=0}^{\infty} E\left(e^{iuC_1}\ldots e^{iuC_s}e^{i\xi T}|S = s\right)P(S = s). \tag{4.6.9}$$

We shall prove that the random variables S, C_1, ..., C_s, T are independent. The random variables of the sequence $\{C_s, s = 1, 2, \ldots\}$ are equally distributed with the random variable C having characteristic function $F_C(c)$.

The independence of $\{C_s, s = 1, 2, \ldots\}$, S, $\{X_n, n = 1, 2, \ldots\}$ and N implies the independence of N, $\{X_n, n = 1, 2, \ldots\}$.

Hence (2.7.8), (4.6.3) and (4.6.4) imply that the distribution function $F_T(t)$ of the random variable $T = \min(X_1, X_2, \ldots, X_N)$ is

$$F_T(t) = 1 - P_N(1 - F_X(t)). \tag{4.6.10}$$

We consider the vector $(S, C_1, \ldots, C_s, T)$ of random variables S, C_1, …, C_s, T and $F_{S,C_1,\ldots,C_s,T}(s, c_1, \ldots, c_s, t)$ be the distribution function of that vector. Let $F_S(s)$ be the distribution function of the random variable S.

The proof of independence of the random variables S, C_1, …, C_s, T requires the proof of the relationship

$$F_{S,C_1,\ldots,C_s,T}(s, c_1, \ldots, c_s, t) = F_S(s)F_{C_1}(c_1)\ldots F_{C_s}(c_s)F_T(t)$$

or equivalently the proof of the relationship

$$F_{S,C_1,\ldots,C_s,T}(s, c_1, \ldots, c_s, t) = F_S(s)F_C(c_1)\ldots F_C(c_s)F_T(t)$$

We have

$$\begin{aligned} P(S \le s, C_1 \le c_1, \ldots, C_s \le c_s) = {} & P(S \le s, C_1 \le c_1, \ldots, C_s \le c_s, T \le t) \\ & + P(S \le s, C_1 \le c_1, \ldots, C_s \le c_s, T > t). \end{aligned} \tag{4.6.11}$$

From (4.6.11) it follows that

$$\begin{aligned} P(S \le s, C_1 \le c_1, \ldots, C_s \le c_s T \le t) = {} & P(S \le s, C_1 \le c_1, \ldots, C_s \le c_s) \\ & - P(S \le s, C_1 \le c_1, \ldots, C_s \le c_s, T > t). \end{aligned} \tag{4.6.12}$$

Since the random variable S is independent of the sequence of the random variables $\{C_s, s = 1, 2, \ldots\}$ then the random variables S, C_1, …, C_s, are independent. Hence we get that

$$P(S \le s, C_1 \le c_1, \ldots, C_s \le c_s) = P(S \le s)P(C_1 \le c_1)\ldots P(C_s \le c_s)$$

or equivalently we get that

$$P(S \le s, C_1 \le c_1, \ldots, C_s \le c_s) = F_S(s)F_{C_1}(c_1)\ldots F_{C_s}(c_s). \tag{4.6.13}$$

From (4.6.13) it follows that

$$P(S \le s, C_1 \le c_1, \ldots, C_s \le c_s) = F_S(s)F_C(c_1)\ldots F_C(c_s). \tag{4.6.14}$$

Moreover (4.6.12) and (4.6.14) imply that

$$P(S \le s, C_1 \le c_1, \ldots, C_s \le c_s, T \le t) = F_S(s)F_C(c_1)\ldots F_C(c_s) - P(S \le s, C_1 \le c_1, \ldots, C_s \le c_s, T > t). \quad (4.6.15)$$

Hence (4.6.15) implies that

$$P(S \le s, C_1 \le c_1, \ldots, C_S \le c_s, T \le t) = F_S(s)F_C(c_1)\ldots F_C(c_s) - \sum_{n=1}^{\infty} P(S \le s, C_1 \le c_1, \ldots, C_s \le c_s, T > t|N = n)P(N = n). \quad (4.6.16)$$

From (4.6.16) it follows that

$$P(S \le s, C_1 \le c_1, \ldots, C_s \le c_s, T \le t) = F_S(s)F_C(c_1)\ldots F_C(c_s) - \sum_{n=1}^{\infty} P(S \le s, C_1 \le c_1, \ldots, C_s \le c_s, \min(X_1, \ldots, X_N) > t|N = n)P(N = n). \quad (4.6.17)$$

From (4.6.17) we get that

$$P(S \le s, C_1 \le c_1, \ldots, C_s \le c_s, T \le t) = F_S(s)F_C(c_1)\ldots F_C(c_s) - \sum_{n=1}^{\infty} P[S \le s, C_1 \le c_1, \ldots, C_s \le c_s, \min(X_1, \ldots, X_n) > t|N = n]P(N = n) \quad (4.6.18)$$

Hence (4.6.18) implies that

$$P(S \le s, C_1 \le c_1, \ldots, C_s \le c_s, T \le t) = F_S(s)F_C(c_1)\ldots F_C(c_s) - \sum_{n=1}^{\infty} P(S \le s, C_1 \le c_1, \ldots, C_s \le c_s, X_1 > t, \ldots, X_n > t|N = n)P(N = n). \quad (4.6.19)$$

Since N, $\{X_n, n = 1, 2, \ldots\}$, S and $\{C_s, s = 1, 2, \ldots\}$ are independent then the random variables N, S, C_1, …, C_s, X_1, …, X_n are independent. Hence (4.6.19) has the form

$$P(S \le s, C_1 \le c_1, \ldots, C_s \le c_s, T \le t) = F_S(s)F_C(c_1)\ldots F_C(c_s) - \sum_{n=1}^{\infty} P(S \le s, C_1 \le c_1, \ldots, C_s \le c_s, X_1 > t, \ldots, X_n > t)P(N = n). \tag{4.6.20}$$

The independence of random variables N, S, C_1, …, C_s, X_1, …, X_n implies the independence of the random variables S, C_1, …, C_s, X_1, …, X_n

Hence (4.6.20) has the form

$$P(S \le s, C_1 \le c_1, \ldots, C_s \le c_s, T \le t) = F_S(s)F_C(c_1)\ldots F_C(c_s) - \sum_{n=1}^{\infty} P(S \le s)P(C_1 \le c_1)\ldots P(C_s \le c_s) \times P(X_1 > t)\ldots P(X_n > t)P(N = n)$$

or equivalently the form

$$P(S \le s, C_1 \le c_1, \ldots, C_s \le c_s, T \le t) = F_S(s)F_C(c_1)\ldots F_C(c_s) - P(S \le s)P(C_1 \le c_1)\ldots P(C_s \le c_s) \times \sum_{n=1}^{\infty} P(X_1 > t)\ldots P(X_n > t)P(N = n). \tag{4.6.21}$$

From (4.6.21) it follows that

$$P(S \le s, C_1 \le c_1, \ldots, C_s \le c_s, T \le t) = F_S(s)F_C(c_1)\ldots F_C(c_s) - F_S(s)F_C(c_1)\ldots F_C(c_s) \times \sum_{n=1}^{\infty} (1 - F_X(t))^n P(N = n). \tag{4.6.22}$$

Hence (4.6.22) implies that

$$P(S \le s, C_1 \le c_1, \ldots, C_s \le c_s, T \le t) = F_S(s)F_C(c_1)\ldots F_C(c_s) - F_S(s)F_C(c_1)\ldots F_C(c_s)P_N(1 - F_X(t)). \tag{4.6.23}$$

From (4.6.23) it follows that

$$P(S \leq s, C_1 \leq c_1, \ldots, C_s \leq c_s, T \leq t) = F_S(s)F_C(c_1)\ldots F_C(c_s)[1 - P_N(1 - F_X(t))]. \tag{4.6.24}$$

Moreover (4.6.10) and (4.6.24) imply that

$$P(S \leq s, C_1 \leq c_1, \ldots, C_s \leq c_s, T \leq t) = F_S(s)F_C(c_1)\ldots F_C(c_s)F_T(t) \tag{4.6.25}$$

Since

$$P(S \leq s, C_1 \leq c_1, \ldots, C_s \leq c_s, T \leq t) = F_{S,C_1,\ldots,C_s,T}(s, c_1, \ldots, c_s, t)$$

then (4.6.25) has the form

$$F_{S,C_1,\ldots,C_s,T}(s, c_1, \ldots, c_s, t) = F_S(s)F_C(c_1)\ldots F_C(c_s)F_T(t).$$

Hence the random variables S, C_1, …, C_s, T are independent. That implies the independence of the random variables S, e^{iuC_1}, …, e^{iuC_s}, $e^{i\xi T}$.
Hence (4.6.9) has the form

$$\varphi_{L,T}(u, \xi) = \sum_{s=0}^{\infty} E(e^{iuC_1}\ldots e^{iuC_s}e^{i\xi T})P(S = s). \tag{4.6.26}$$

Since the independence of the random variables S, e^{iuC_1}, …, e^{iuC_s}, $e^{i\xi T}$ implies the independence of the random variables e^{iuC_1}, …, e^{iuC_s}, $e^{i\xi T}$ then (4.6.26) has the form

$$\varphi_{L,T}(u, \xi) = \sum_{s=0}^{\infty} E(e^{iuC_1})\ldots E(e^{iuC_s})E(e^{i\xi T})P(S = s). \tag{4.6.27}$$

Since the random variables of the sequence $\{C_s, s = 1, 2, \ldots\}$ are equally distributed with the random variable C having characteristic function $\varphi_C(u)$ and

$$\varphi_T(\xi) = E(e^{i\xi T})$$

is the characteristic function of the random variable $T = \min(X_1, X_2, \ldots, X_N)$ then (4.6.27) has the form

$$\varphi_{L,T}(u, \xi) = \varphi_T(\xi) \sum_{s=0}^{\infty} \varphi_C^s(u)P(S = s). \tag{4.6.28}$$

From (4.6.28) it follows that

$$\varphi_{L,T}(u,\xi) = \varphi_T(\xi)P_S(\varphi_C(u)). \tag{4.6.29}$$

Since

$$\varphi_L(u) = P_S(\varphi_C(u))$$

then (4.6.29) has the form

$$\varphi_{L,T}(u,\xi) = \varphi_L(u)\varphi_T(\xi).$$

Hence (4.6.5) implies the independence of the random variables $L = C_1 + C_2 + \cdots + C_S$ and $T = \min(X_1, X_2, \ldots, X_N)$.

We consider the random variable $W = \mathrm{e}^{-rT}$.

The independence of the random variables $L = C_1 + C_2 + \cdots + C_S$ and $T = \min(X_1, X_2, \ldots, X_N)$ implies the independence of the random variables $L = C_1 + C_2 + \cdots + C_S$ and $W = \exp[-r\min(X_1, X_2, \ldots, X_N)]$.

The evaluation of the characteristic function $\varphi_Y(u)$ of the stochastic model

$$Y = Le^{-rT}$$

or equivalently the stochastic model $Y = LW$ requires the evaluation of the distribution function of the random variable $W = \mathrm{e}^{-rT}$.

From (4.6.10) we get that

$$F_T(t) = 1 - P_N(1 - F_X(t))$$

is the distribution function of the random variable $T = \min(X_1, X_2, \ldots, X_N)$.

We have

$$F_W(w) = P(W \leq w)$$

or equivalently we have

$$F_W(w) = P\left(e^{-rT} \leq w\right). \tag{4.6.30}$$

From (4.6.30) it follows that

$$F_W(w) = P(-rT \leq \log w)$$

or equivalently

$$F_W(w) = P\left(T \geq -\frac{1}{r}\log w\right). \tag{4.6.31}$$

From (4.6.31) it follows that

$$F_W(w) = 1 - P\left(T \leq -\frac{1}{r}\log w\right). \tag{4.6.32}$$

Since

$$P(T \leq t) = F_T(t)$$

and

$$F_T(t) = 1 - P_N(1 - F_X(t))$$

then (4.6.32) implies that the distribution function $F_W(w)$ of the random variable $W = \mathrm{e}^{-rT}$ is

$$F_W(w) = P_N\left(1 - F_X\left(-\frac{1}{r}\log w\right)\right), \quad 0<w<1. \tag{4.6.33}$$

The evaluation of the characteristic function $\varphi_Y(u)$ of the stochastic model

$$Y = L\mathrm{e}^{-rT}$$

or equivalently the stochastic model $Y = LW$ is implemented as follows. We have

$$\varphi_Y(u) = E\left(e^{iuLW}\right)$$

or equivalently we have

$$\varphi_Y(u) = E\left(E\left(e^{iuLW}|W\right)\right). \tag{4.6.34}$$

From (4.6.34) it follows that

$$\varphi_Y(u) = \int_0^1 E\left(e^{iuLW}|W = w\right)dF_W(w). \tag{4.6.35}$$

Hence (4.6.35) implies that

$$\varphi_Y(u) = \int_0^1 E\left(e^{iuwL}|W = w\right)dF_W(w). \tag{4.6.36}$$

Since the independence of the random variables $L = C_1 + C_2 + \cdots + C_S$ and $T = \min(X_1, X_2, \ldots, X_N)$ implies the independence of the random variables $L =$

$C_1 + C_2 + \cdots + C_S$ and $W = \exp[-r\min(X_1, X_2, \ldots, X_N)]$ then (4.6.36) implies that

$$\varphi_Y(u) = \int_0^1 E(e^{iuwL})dF_W(w). \tag{4.6.37}$$

Since the characteristic function of the random variable $L = C_1 + C_2 + \cdots + C_S$ is

$$\varphi_L(u) = P_S(\varphi_C(u)) \tag{4.6.38}$$

and (4.6.33) implies that the distribution function of the random variable $W = \exp[-r\min(X_1, X_2, \ldots, X_N)]$ is

$$F_W(w) = P_N\left(1 - F_X\left(-\frac{1}{r}\log w\right)\right) \tag{4.6.39}$$

then (4.6.37), (4.6.38) and (4.6.39) imply that the characteristic function of the stochastic model

$$Y = Le^{-rT}$$

is

$$\varphi_Y(u) = \int_0^1 P_S(\varphi_C(uw))dP_N\left(1 - F_X\left(-\frac{1}{r}\log w\right)\right). \tag{4.6.40}$$

For embedding the characteristic function (4.6.40) in the class of characteristic functions corresponding to a-unimodal probability distributions we work as follows.

We suppose that the random variable X follows the exponential distribution with distribution function

$$F_X(x) = 1 - e^{-\mu x},$$

the random variable N follows the degenerate distribution with probability generating function

$$P_N(z) = z^n,$$

and we set $a = \mu n/r$, then (4.6.40) implies that the characteristic function of the stochastic model

$$Y = Le^{-rT}$$

has the form

$$\varphi_Y(u) = a\int_0^1 P_S(\varphi_C(uw))w^{a-1}dw \tag{4.6.41}$$

which means that the characteristic (4.6.41) corresponds to α-unimodal probability distribution.

An application of the stochastic model

$$Y = Le^{-rT},$$

where $L = C_1 + C_2 + \cdots + C_S$ and $T = \min(X_1, X_2, \ldots, X_N)$ in considering an organization under conditions of competing risks is the following.

We suppose that the random variable N denotes the number of risks threatening an organization at the time point 0 and the random variable X_n denotes the occurrence time of the nth risk then the random variable $T = \min(X_1, X_2, \ldots, X_N)$ denotes the minimum time of occurrence of a risk. We suppose that the random variable S denotes the number of different damages due to the risk occurring at the time point $T = \min(X_1, X_2, \ldots, X_N)$ and the random variable C_s denotes the size of the sth damage of that risk then the random variable $L = C_1 + C_2 + \cdots + C_S$ denotes the total size of the S damages due to the risk occurring at the time point $T - \min(X_1, X_2, \ldots, X_N)$.

Hence the stochastic model

$$Y = Le^{-rT}$$

denotes the present value at the time point 0 of the total size $L = C_1 + C_2 + \cdots + C_S$ of the S damages due to the risk occurring at the time point $T = \min(X_1, X_2, \ldots, X_N)$.

From the fact that one risk occurs in the time interval $[0, T]$ and the occurrence of that risk arises at the time point $T = \min(X_1, X_2, \ldots, X_N)$ we get that the stochastic model

$$Y = Le^{-rT}$$

is particularly useful for developing and implementing a plan for financing the damage $L = C_1 + C_2 + \cdots + C_S$ due to the risk occurring at the time point $T = \min(X_1, X_2, \ldots, X_N)$.

Two forms of such a financing plan are the following.

The first form of the financing plan can be creation of a reserve of size

$$Y = Le^{-rT}$$

at the time point 0 which is continuously compounded in the time interval $[0, T]$ with force of interest r.

The second form of the financing plan can be the creation of a continuous uniform cash flow starting at the time 0 with force of interest r, duration $T = \min(X_1, X_2, \ldots, X_N)$ and future value $L = C_1 + C_2 + \cdots + C_S$.

Another application of the stochastic model

$$Y = Le^{-rT}$$

in considering an organization under conditions of competing risks arises if the random sum $L = C_1 + C_2 + \cdots + C_S$ is interpreted in the following way. We suppose that the random variable N denotes the number of risks threatening an organization at the time point 0 and the random variable X_n denotes the occurrence time of the nth risk then the random variable $T = \min(X_1, X_2, \ldots, X_N)$ denotes the minimum time of occurrence of a risk. We suppose that the random variable S denotes the number of banks participating in the share capital of the organization at the time point $T = \min(X_1, X_2, \ldots, X_N)$.

The random variable C_s denotes the size of the share capital of the organization belonging to the sth bank at the time point $T = \min(X_1, X_2, \ldots, X_N)$.

Hence the random variable $L = C_1 + C_2 + \cdots + C_S$ denotes the size of the share capital of the organization belonging to the S banks at the time point $T = \min(X_1, X_2, \ldots, X_N)$.

In this case the stochastic

$$Y = Le^{-rT}$$

denotes the present value at the time point 0 of the size $L = C_1 + C_2 + \cdots + C_S$ of the share capital of the organization belonging to the S banks at the time point $T = \min(X_1, X_2, \ldots, X_N)$.

The presence of the random variables $L = C_1 + C_2 + \cdots + C_S$ and $T = \min(X_1, X_2, \ldots, X_N)$ in the stochastic model

$$Y = Le^{-rT}$$

makes the above application of that stochastic model particularly useful for investigating the capital structure of an organization under conditions of competing risks.

An application of the stochastic model

$$Y = Le^{-rT},$$

where $L = C_1 + C_2 + \cdots + C_S$ and $T = \min(X_1, X_2, \ldots, X_N)$ in banking is the following.

We suppose that the random variable N denotes the number of loans in the portfolio of loans of a bank at the time point 0 and the random variable X_n denotes the expiration time point of the nth loan then the random variable $T = \min(X_1, X_2, \ldots, X_N)$ denotes the minimum expiration time point for the loans of the portfolio of loans of the bank. We suppose that the bank participates in the share capital of the S firms at the time point $T = \min(X_1, X_2, \ldots, X_N)$.

The random variable C_s denotes the size of the share capital of the sth firm which belongs to the bank at the time point $T = \min(X_1, X_2, \ldots, X_N)$.

Hence the random variable $L = C_1 + C_2 + \cdots + C_S$ denotes the size of the share capital of the S firms which belongs to the bank at the time point $T = \min (X_1, X_2, \ldots, X_N)$

In this case the stochastic model

$$Y = Le^{-rT}$$

denotes the present value at the time point 0 of the size $L = C_1 + C_2 + \cdots + C_S$ of the share capital of the S firms which belongs to the bank at the time point $T = \min(X_1, X_2, \ldots, X_N)$.

Another application of the stochastic model

$$Y = Le^{-rT}$$

in banking arises if the random sum $L = C_1 + C_2 + \cdots + C_S$ is interpreted in the following way. We suppose that the random variable N denotes the number of loans in the portfolio of loans of a bank at the time point 0 and the random variable X_n denotes the expiration time point of the nth loan then the random variable $T = \min(X_1, X_2, \ldots, X_N)$ denotes the minimum expiration time point for the loans of the portfolio of loans of the bank. We suppose that the random variable S denotes the number of time deposits in the bank at the time point $T = \min(X_1, X_2, \ldots, X_N)$.

The random variable C_s denotes the size of the sth time deposit at the time point $T = \min(X_1, X_2, \ldots, X_N)$.

Hence the random variable $L = C_1 + C_2 + \cdots + C_S$ denotes the size of the S time deposits in the bank at the time point $T = \min(X_1, X_2, \ldots, X_N)$.

In this case the stochastic model

$$Y = Le^{-rT}$$

denotes the present value at the time point 0 of the size $L = C_1 + C_2 + \cdots + C_S$ of the S time deposits in the bank at the time point $T = \min(X_1, X_2, \ldots, X_N)$.

An application of the stochastic model

$$Y = Le^{-rT}$$

where $L = C_1 + C_2 + \cdots + C_S$ and $T = \min(X_1, X_2, \ldots, X_N)$ in industrial activities is the following.

We suppose that the random variable N denotes the number of orders that an industrial firm implements at the time point 0 and the random variable X_n denotes the time required for completing the implementation of the nth order then the random variable $T = \min(X_1, X_2, \ldots, X_N)$ denotes the minimum time required for completing the implementation of some order. We suppose that the random variable S denotes the number of investments in the portfolio of investments of the industrial firm at the time point $T = \min(X_1, X_2, \ldots, X_N)$.

The random variable C_s denotes the market value of the sth investment in the portfolio of investments of the industrial firm at the time point $T = \min(X_1, X_2, \ldots, X_N)$.

Hence the random variable $L = C_1 + C_2 + \cdots + C_S$ denotes the market value of the *sth* investments in the portfolio of investments of the industrial firm at the time point $T = \min(X_1, X_2, \ldots, X_N)$.

In this case the stochastic model

$$Y = Le^{-rT}$$

denotes the present value at the time point 0 of the market value $L = C_1 + C_2 + \cdots + C_S$ of the *sth* investments in the portfolio of investments of the industrial firm at the time point $T = \min(X_1, X_2, \ldots, X_N)$.

Another application of the stochastic model

$$Y = Le^{-rT}$$

in industrial activities arises if the random sum $L = C_1 + C_2 + \cdots + C_S$ is interpreted as follows. We suppose that the random variable N denotes the number of orders that an industrial firm implements at the time point 0 and the random variable X_n denotes the time required for completing the implementation of the nth order then the random variable $T = \min(X_1, X_2, \ldots, X_N)$ denotes the minimum time required for completing the implementation of some order.

We suppose that the random variable S denotes the number of loans in the portfolio of loans of the industrial firm at the time point $T = \min(X_1, X_2, \ldots, X_N)$.

The random variable C_s denotes the size of the sth loan in the portfolio of loans of the industrial firm at the time point $T = \min(X_1, X_2, \ldots, X_N)$.

Hence the random variable $L = C_1 + C_2 + \cdots + C_S$ denotes the size of the S of loans in the portfolio of loans of the industrial firm at the time point $T = \min(X_1, X_2, \ldots, X_N)$.

In this case the stochastic model

$$Y = Le^{-rT}$$

denotes the present value at the time point 0 of the size $L = C_1 + C_2 + \cdots + C_S$ of the S loans of the loan portfolio of the industrial firm at the minimum required time for completing the implementation of the order $T = \min(X_1, X_2, \ldots, X_N)$.

An interesting modification of the stochastic model

$$Y = Le^{-rT}$$

where $L = C_1 + C_2 + \cdots + C_S$ and $T = \min(X_1, X_2, \ldots, X_N)$ is obtained as follows.

Let N be a discrete random variable with values in the set $\mathbf{N} = \{1, 2, \ldots\}$ and probability generating function $P_N(z)$.

We suppose that $\{X_n, n = 1, 2, \ldots\}$ be a sequence of continuous, positive, independent, and identically distributed random variables. The random variables of the sequence are equally distributed with the random variable X having distribution function $F_X(x)$ and we set $T = \min(X_1, X_2, \ldots, X_N)$.

Let Π be a continuous and positive random variable with characteristic function $\varphi_\Pi(u)$ and U be a continuous and positive random variable with characteristic function $\varphi_U(u)$.

We consider the positive real number r and we set

$$V = U + \Pi \exp[-r \min(X_1, X_2, \ldots, X_N)]$$

or equivalently

$$V = U + \Pi e^{-rT}.$$

The establishment of properties and applications in risk management of the stochastic model

$$V = U + \Pi e^{-rT}$$

is of particular practical and theoretical interest. An interpretation in the area of continuous discounting of the above stochastic model is the following.

We suppose that any random variable of the sequence $\{X_n, n = 1, 2, \ldots\}$ denotes time and the random variable Π denotes a cash flow, the random variable U denotes a cash flow and the positive real number r denotes force of interest. In this case the stochastic model

$$V = U + \Pi e^{-rT}$$

denotes the sum of the cash flow U corresponding at the time point 0 and the present value at the time point 0 of the cash flow Π corresponding at the time point $T = \min(X_1, X_2, \ldots, X_N)$.

The following result establishes sufficient conditions for evaluating the characteristic function $\varphi_V(u)$ of the stochastic model

$$V = U + \Pi e^{-rT}.$$

□

Theorem 4.6.2 *Let N be a discrete random variable with values in the set $\mathbf{N} = \{1, 2, \ldots\}$ and probability generating function $P_N(z)$.*

We suppose that $\{X_n, n = 1, 2, \ldots\}$ be a sequence of continuous, positive, independent, and identically distributed random variables. The random variables of the sequence are equally distributed with the random variable X having distribution function $F_X(x)$ and we set $T = \min(X_1, X_2, \ldots, X_N)$.

Let Π be continuous and positive random variable with characteristic function $\varphi_\Pi(u)$ and U be continuous and positive random variable with characteristic function $\varphi_U(u)$.

We consider the positive real number r and we set

$$V = U + \Pi \mathrm{e}^{-rT}.$$

If N, $\{X_n, n = 1, 2, \ldots\}$, Π and U are independent then the characteristic function of the stochastic model

$$V = U + \Pi \mathrm{e}^{-rT}$$

is

$$\varphi_V(u) = \varphi_U(u) \int_0^1 \varphi_\Pi(uw) dP_N\left(1 - F_X\left(-\frac{1}{r}\log w\right)\right).$$

Proof The independence of N, $\{X_n, n = 1, 2, \ldots\}$, Π and U implies the independence of the random variable N and the sequence of random variables $\{X_n, n = 1, 2, \ldots\}$.

Hence the random variable $T = \min(X_1, X_2, \ldots, X_N)$ has distribution function

$$F_T(t) = 1 - P_N(1 - F_X(t))$$

and the random variable $W = \mathrm{e}^{-rT}$ has distribution function

$$F_W(w) = P_N\left(1 - F_X\left(-\frac{1}{r}\log w\right)\right), \quad 0 < w < 1.$$

Let $F_U(\upsilon)$ be the distribution function of the random variable U, $F_\Pi(\pi)$ the distribution function of the random variable Π,

$$F_T(t) = 1 - P_N(1 - F_X(t))$$

the distribution function of the random variable $T = \min(X_1, X_2, \ldots, X_N)$ and $F_{U,\Pi,T}(\upsilon, \pi, t)$ the distribution function of the vector (U, Π, T).

The establishment of the relationship

$$F_{U,\Pi,T}(\upsilon,\pi,t) = F_U(\upsilon)F_\Pi(\pi)F_T(t)$$

implies the independence of the random variables U, Π, T.

We have

$$F_{U,\Pi,T}(\upsilon,\pi,t) = P(U \le \upsilon, \Pi \le \pi, T \le t) \tag{4.6.42}$$

and

$$P(U \le \upsilon, \Pi \le \pi) = P(U \le \upsilon, \Pi \le \pi, T \le t) + P(U \le \upsilon, \Pi \le \pi, T > t). \tag{4.6.43}$$

From (4.6.43) it follows that

$$P(U \le \upsilon, \Pi \le \pi, T \le t) = P(U \le \upsilon, \Pi \le \pi) - P(U \le \upsilon, \Pi \le \pi, T > t). \tag{4.6.44}$$

Hence (4.6.42) and (4.6.44) imply that

$$F_{U,\Pi,T}(\upsilon,\pi,t) = P(U \le \upsilon, \Pi \le \pi) - \sum_{n=1}^{\infty} P(U \le \upsilon, \Pi \le \pi, T > t | N = n)P(N = n). \tag{4.6.45}$$

Moreover (4.6.45) implies that

$$F_{U,\Pi,T}(\upsilon,\pi,t) = P(U \le \upsilon, \Pi \le \pi) - \sum_{n=1}^{\infty} P(U \le \upsilon, \Pi \le \pi, \min(X_1, X_2, \ldots, X_n) > t | N = n)P(N = n). \tag{4.6.46}$$

Hence (4.6.46) implies that

$$F_{U,\Pi,T}(\upsilon,\pi,t) = P(U \le \upsilon, \Pi \le \pi) - \sum_{n=1}^{\infty} P(U \le \upsilon, \Pi \le \pi, X_1 > t, \ldots, X_n > t | N = n)P(N = n). \tag{4.6.47}$$

The independence of N, $\{X_n, n = 1, 2, \ldots\}$, Π and U implies the independence of N, X_1, …, X_n, Π, U.

Hence (4.6.47) has the form

$$F_{U,\Pi,T}(\upsilon,\pi,t) = P(U \le \upsilon, \Pi \le \pi) - \sum_{n=1}^{\infty} P(U \le \upsilon, \Pi \le \pi, X_1 > t, \ldots, X_n > t)P(N = n). \tag{4.6.48}$$

The independence of the random variables N, X_1, …, X_n, Π, U implies the independence of the random variables X_1, …, X_n, Π, U and the independence of the random variables U, Π.

Hence (4.6.48) has the form

$$F_{U,\Pi,T}(\upsilon,\pi,t) = P(U \le \upsilon)P(\Pi \le \pi) - \sum_{n=1}^{\infty} P(U \le \upsilon)P(\Pi \le \pi) \times P(X_1 > t)\ldots P(X_n > t)P(N = n)$$

or equivalently the form

$$F_{U,\Pi,T}(\upsilon,\pi,t) = F_U(\upsilon)F_\Pi(\pi) - F_U(\upsilon)F_\Pi(\pi)\sum_{n=1}^{\infty}(1 - F_X(t))^n P(N = n). \tag{4.6.49}$$

From (4.6.49) implies that

$$F_{U,\Pi,T}(\upsilon,\pi,t) = F_U(\upsilon)F_\Pi(\pi)(1 - P_N(1 - F_X(t)))$$

or equivalently

$$F_{U,\Pi,T}(\upsilon,\pi,t) = F_U(\upsilon)F_\Pi(\pi)F_T(t). \tag{4.6.50}$$

Hence (4.6.50) implies that the random variables U, Π, T are independent. The independence of the above random variables implies the independence of the random variables U, Π, $W = \mathrm{e}^{-rT}$.

The independence of the random variables U, Π, $W = e^{-rT}$ implies the independence of the random variables Π, W.

If $\varphi_{\Pi W}(u)$ is the characteristic function of the random variableί ΠW then we get that

$$\varphi_{\Pi W}(u) = E\left(e^{iu\Pi W}\right)$$

or equivalently we get that

$$\varphi_{\Pi W}(u) = E\left(e^{iu\Pi W}|W\right). \tag{4.6.51}$$

From (4.6.51) we get that

$$\varphi_{\Pi W}(u) = \int_0^1 E\left(e^{iu\Pi W}|W = w\right)dF_W(w)$$

or equivalently we get that

$$\varphi_{\Pi W}(u) = \int_0^1 E(e^{iw\Pi}|W = w)dP_N\left(1 - F_X\left(-\frac{1}{r}\log w\right)\right). \tag{4.6.52}$$

Since the random variables Π, W are independent then (4.6.52) has the form

$$\varphi_{\Pi W}(u) = \int_0^1 \varphi_\Pi(uw)dP_N\left(1 - F_X\left(-\frac{1}{r}\log w\right)\right).$$

If $\varphi_V(u)$ is the characteristic function of the stochastic model

$$V = U + \Pi e^{-rT}$$

or equivalently the stochastic model

$$V = U + \Pi W$$

then we get that

$$\varphi_V(u) = E\left(e^{iu(U+\Pi W)}\right). \tag{4.6.53}$$

Hence (4.6.53) implies that

$$\varphi_V(u) = E\left(E\left(e^{iu(U+\Pi W)}\right)|W\right). \tag{4.6.54}$$

From (4.6.54) it follows that

$$\varphi_V(u) = \int_0^1 E\left(e^{iu(U+\Pi W)}|W = w\right)dF_W(w) \tag{4.6.55}$$

and (4.6.55) implies that

$$\varphi_V(u) = \int_0^1 E(e^{iuU+iu\Pi W}|W = w)dP_N\left(1 - F_X\left(-\frac{1}{r}\log w\right)\right). \tag{4.6.56}$$

Hence (4.6.56) implies that

$$\varphi_V(u) = \int_0^1 E(e^{iuU+iuw\Pi}|W = w)dP_N\left(1 - F_X\left(-\frac{1}{r}\log w\right)\right). \tag{4.6.57}$$

Since the random variables U, Π, $W = \mathrm{e}^{-rT}$ are independent then (4.6.57) has the form

$$\varphi_V(u) = \int_0^1 E(e^{iuU+iuw\Pi})dP_N\left(1 - F_X\left(-\frac{1}{r}\log w\right)\right). \tag{4.6.58}$$

Since the independence of the random variables U, Π, $W = \mathrm{e}^{-rT}$ implies the independence of the random variables U, Π then (4.6.58) has the form

$$\varphi_V(u) = \int_0^1 E(e^{iuU})E(e^{iuw\Pi})dP_N\left(1 - F_X\left(-\frac{1}{r}\log w\right)\right). \tag{4.6.59}$$

From (4.6.59) it follows that

$$\varphi_V(u) = E(e^{iuU})\int_0^1 E(e^{iuw\Pi})dP_N\left(1 - F_X\left(-\frac{1}{r}\log w\right)\right). \tag{4.6.60}$$

Since

$$\varphi_U(u) = E(e^{iuU})$$

is the characteristic function of the random variable U and

$$\varphi_\Pi(u) = E(e^{iu\Pi})$$

is the characteristic function of the random variable Π then (4.6.60) implies that the characteristic function of the stochastic model

$$V = U + \Pi\mathrm{e}^{-rT},$$

where $T = \min(X_1, X_2, \ldots, X_N)$, is

$$\varphi_V(u) = \varphi_U(u)\int_0^1 \varphi_\Pi(uw)dP_N\left(1 - F_X\left(-\frac{1}{r}\log w\right)\right)$$

or equivalently

$$\varphi_V(u) = \varphi_U(u)\varphi_{\Pi W}(u).$$

An application of the stochastic model

$$V = U + \Pi e^{-rT},$$

where $T = \min(X_1, X_2, \ldots, X_N)$, in risk management is the following. We suppose that the random variable N denotes the number of risks threatening a firm at the time point 0 and the random variable X_n denotes the occurrence time of the nth risk then the random variable $T = \min(X_1, X_2, \ldots, X_N)$ denotes the minimum risk occurrence time. We suppose that the random variable U denotes the size of a loan that the firm undertakes at the time point 0 and the random variable Π denotes the size of another loan that the firm undertakes at the time point $T = \min(X_1, X_2, \ldots, X_N)$.

The random variable Πe^{-rT} denotes the present value of the size Π of the loan that the firm undertakes at the time point $T = \min(X_1, X_2, \ldots, X_N)$.

Hence the stochastic model

$$V = U + \Pi e^{-rT}$$

denotes the total loan obligation of the firm at the time point 0. The following result concentrates on the establishment of the characteristic function $\varphi_V(u)$ of a special case of the stochastic model

$$V = U + \Pi e^{-rT}.$$

□

Theorem 4.6.3 *Let N be a discrete random variable following the degenerate distribution with probability generating function*

$$P_N(z) = z^n.$$

We suppose that $\{X_n, n = 1, 2, \ldots\}$ is a sequence of continuous, positive, independent, and identically distributed random variables. The random variables of the sequence are equally distributed with the random variable X following the exponential distribution with distribution function

$$F_X(x) = 1 - e^{-\mu x}, \quad x > 0 \quad \mu > 0$$

and we set $T = \min(X_1, X_2, \ldots, X_N)$.

Let Π be a continuous and positive random variable with characteristic function $\varphi_\Pi(u)$ and finite mean. Let U be a continuous and positive random variable with characteristic function $\varphi_U(u)$ and finite mean. We consider the stochastic model

$$V = U + \Pi e^{-rT}$$

where $r > 0$.

We suppose that N, $\{X_n, n = 1, 2, \ldots\}$, Π and U are independent. The characteristic function of the stochastic model

$$V = U + \Pi e^{-rT}$$

is

$$\varphi_V(u) = \varphi_U(u)\exp\left(a\int_0^u \frac{\varphi_U(y)-1}{y}dy\right),$$

where $a = n\mu/r$ *if, and only, if* $V \overset{d}{=} \Pi$.

Proof From Theorem 4.6.2 it follows that the characteristic function of the stochastic model

$$V = U + \Pi e^{-rT}$$

is

$$\varphi_V(u) = \varphi_U(u)\int_0^1 \varphi_\Pi(uw)dP_N\left(1 - F_X\left(-\frac{1}{r}\log w\right)\right). \tag{4.6.61}$$

Since the probability generating function of the random variable N is

$$P_N(z) = z^n$$

and the distribution function of the random variable X is

$$F_X(x) = 1 - e^{-\mu x}, \quad x > 0 \quad \mu > 0$$

then the distribution function of the random variable $T = \min(X_1, X_2, \ldots, X_N)$ is

$$F_T(t) = 1 - e^{-n\mu t}$$

and the distribution function of the random variable $W = e^{-rT}$ is

$$F_W(w) = w^a, \quad 0 < w < 1$$

where

$$a = n\mu/r.$$

Hence (4.6.61) has the form

$$\varphi_V(u) = \varphi_U(u)a\int_0^1 \varphi_\Pi(uw)w^{a-1}dw$$

or equivalently the form

$$\varphi_V(u) = \varphi_U(u) \cdot \frac{a}{u^a} \int_0^u \varphi_\Pi(w) w^{a-1} dw. \tag{4.6.62}$$

We suppose that the random variables V, Π are equally distributed then we get that

$$\varphi_V(u) = \varphi_\Pi(u). \tag{4.6.63}$$

From (4.6.2) and (4.6.63) it follows that

$$\varphi_V(u) = \varphi_U(u) \cdot \frac{a}{u^a} \int_0^u \varphi_V(w) w^{a-1} dw. \tag{4.6.64}$$

Theorem 4.4.3 implies that the solution of the integral equation (4.6.64) is

$$\varphi_V(u) = \varphi_U(u) \exp\left(a \int_0^u \frac{\varphi_U(y) - 1}{y} dy \right).$$

The inverse is obvious. □

4.7 Time of First Damage for a System Threatened by a Random Number of Risks and Present Value of a Continuous Uniform Cash Flow

Let $\{C_s, s = 1, 2, \ldots\}$ be a sequence of continuous, positive, independent, and identically distributed random variables. We consider the discrete random variable S with values in the set $\mathbf{N}_0 = \{0, 1, 2, \ldots\}$ and probability generating function $P_S(z)$.

We suppose that the random variable S is independent of the sequence of continuous, positive, independent, and identically distributed random variables $\{C_s, s = 1, 2, \ldots\}$ and we set $L = C_1 + C_2 + \cdots + C_S$.

Let $\{X_n, n = 1, 2, \ldots\}$ be a sequence of continuous, positive, independent, and identically distributed random variables. We consider the discrete random variable N with values in the set $\mathbf{N} = \{1, 2, \ldots\}$ and probability generating function $P_N(z)$.

We suppose that the random variable N is independent of the sequence of continuous, positive, independent, and identically distributed random variables $\{X_n, n = 1, 2, \ldots\}$ and we set $T = \min(X_1, X_2, \ldots, X_N)$.

We consider the stochastic model

$$Y = L\frac{1 - e^{-rT}}{r}$$

where r is positive real number. The purpose of the present section is the investigation and applications in risk management of the above stochastic model. An interpretation in the area of continuous discounting of that model is the following.

We suppose that each random variable of the sequence $\{X_n, n = 1, 2, \ldots\}$ denotes time and each random variable of the sequence $\{C_s, s = 1, 2, \ldots\}$ denotes cash flow and the positive real number r denotes force of interest. In this case the stochastic model

$$Y = L\frac{1 - e^{-rT}}{r}$$

denotes the present value at the time point 0 of the continuous uniform cash flow with rate of payment $L = C_1 + C_2 + \cdots + C_S$ and duration $T = \min(X_1, X_2, \ldots, X_N)$.

The following result establishes sufficient conditions for evaluating the characteristic function of the stochastic model

$$Y = L\frac{1 - e^{-rT}}{r}.$$

Theorem 4.7.1 *Let $\{C_s, s = 1, 2, \ldots\}$ be a sequence of continuous, positive, independent, and identically distributed random variables. The random variables of the sequence $\{C_s, s = 1, 2, \ldots\}$ are equally distributed with the random variable C having characteristic function $\varphi_C(u)$.*

We consider the discrete random variable S with values in the set $\mathbf{N}_0 = \{0, 1, 2, \ldots\}$, probability generating function $P_S(z)$ and we set $L = C_1 + C_2 + \cdots + C_S$.

Let $\{X_n, n = 1, 2, \ldots\}$ be a sequence of continuous, positive, independent, and identically distributed random variables. The random variables of the sequence $\{X_n, n = 1, 2, \ldots\}$ are equally distributed with the random variable X having distribution function $F_X(x)$.

We consider the discrete random variable N with values in the set $\mathbf{N} = \{1, 2, \ldots\}$, probability generating function $P_N(z)$ and we set $T = \min(X_1, X_2, \ldots, X_N)$.

We consider the stochastic model

$$Y = L\frac{1 - e^{-rT}}{r}$$

where r is a positive real number. If $\{C_s, s = 1, 2, \ldots\}$, S, $\{X_n, n = 1, 2, \ldots\}$ *and* N *are independent then the characteristic function of the stochastic model*

$$Y = L\frac{1 - e^{-rT}}{r}$$

is

$$\varphi_Y(u) = \int_0^{1/r} P_S(\varphi_C(uw))d\left[1 - P_N\left(1 - F_X\left(-\frac{1}{r}\log(1 - rw)\right)\right)\right]$$

or equivalently

$$\varphi_Y(u) = \int_0^{1} P_S(\varphi_C(^{uw}/_{r}))d\left[1 - P_N\left(1 - F_X\left(-\frac{1}{r}\log(1 - w)\right)\right)\right].$$

Proof We consider the random variable $L = C_1 + C_2 + \cdots + C_S$ and the random variable $T = \min(X_1, X_2, \ldots, X_N)$.

From Theorem 4.6.1 it follows that the random variables $L = C_1 + C_2 + \cdots + C_S$, $T = \min(X_1, X_2, \ldots, X_N)$ are independent where

$$\varphi_L(u) = P_S(\varphi_C(u))$$

is the characteristic function of the random variable $L = C_1 + C_2 + \cdots + C_S$ and $F_T(t) = 1 - P_N(1 - F_X(t))$ is the distribution function of the random variable $T = \min(X_1, X_2, \ldots, X_N)$.

The independence of the random variables $L = C_1 + C_2 + \cdots + C_S$ and $T = \min(X_1, X_2, \ldots, X_N)$ implies the independence of the random variables $L = C_1 + C_2 + \cdots + C_S$,

$$W = \{1 - \exp[-r\min(X_1, X_2, \ldots, X_N)]\}/r.$$

The evaluation of the characteristic function of the stochastic model

$$Y = L\frac{1 - e^{-rT}}{r}$$

or equivalently of the stochastic model $Y = LW$ requires the evaluation of the distribution function $F_W(w)$ of the random variables

$$W = \frac{1 - e^{-rT}}{r}.$$

We have that

$$F_W(w) = P(W \leq w)$$

or equivalently we have that

$$F_W(w) = P\left(\frac{1 - e^{-rT}}{r} \leq w\right). \tag{4.7.1}$$

From (4.7.1) it follows that

$$F_W(w) = P(1 - e^{-rT} \leq rw)$$

or equivalently it follows that

$$F_W(w) = P(-e^{-rT} \leq rw - 1). \tag{4.7.2}$$

Hence (4.7.2) implies that

$$F_W(w) = P(e^{-rT} \geq 1 - rw)$$

or equivalently

$$F_W(w) = P(-rT \geq \log(1 - rw)). \tag{4.7.3}$$

From (4.7.3) it follows that

$$F_W(w) = P\left(T \leq -\frac{1}{r}\log(1 - rw)\right). \tag{4.7.4}$$

Since

$$P(T \leq t) = 1 - P_N(1 - F_X(t))$$

then (4.7.4) implies that the distribution function $F_W(w)$ of the random

$$W = \frac{1 - e^{-rT}}{r}$$

is

$$F_W(w) = 1 - P_N\left(1 - F_X\left(-\frac{1}{r}\log(1 - rw)\right)\right).$$

The independence of the random variables L, W and the distribution function $F_W(w)$ permit the evaluation of the characteristic function $\varphi_Y(u)$ of the stochastic model

$$Y = L\frac{1 - \mathrm{e}^{-rT}}{r}$$

in the following way.

We have

$$\varphi_Y(u) = E\left(e^{iuLW}\right)$$

or equivalently

$$\varphi_Y(u) = E\left(E\left(e^{iuLW}|W\right)\right). \tag{4.7.5}$$

From (4.7.5) it follows that

$$\varphi_Y(u) = \int_0^{1/r} E\left(e^{iuLW}|W = w\right)dF_W(w)$$

or equivalently

$$\varphi_Y(u) = \int_0^{1/r} E\left(e^{iuwL}|W = w\right)dF_W(w). \tag{4.7.6}$$

The independence of the random variables L, W implies that (4.7.6) has the form

$$\varphi_Y(u) = \int_0^{1/r} E\left(e^{iuwL}\right)dF_W(w). \tag{4.7.7}$$

Since

$$\varphi_L(u) = E\left(e^{iuL}\right)$$

then (4.7.7) has the form

$$\varphi_Y(u) = \int_0^{1/r} \varphi_L(uw)dF_W(w). \tag{4.7.8}$$

Since

$$\varphi_L(u) = P_S(\varphi_C(u))$$

and

$$F_W(w) = 1 - P_N\left(1 - F_X\left(-\frac{1}{r}\log(1 - rw)\right)\right)$$

then (4.7.8) has the form

$$\varphi_Y(u) = \int_0^{1/r} P_S(\varphi_C(uw))d\left[1 - P_N\left(1 - F_X\left(-\frac{1}{r}\log(1 - rw)\right)\right)\right]. \quad (4.7.9)$$

From (4.7.9) it follows that

$$\varphi_Y(u) = \int_0^{1} P_S(\varphi_C(^{uw}/_{r}))d\left[1 - P_N\left(1 - F_X\left(-\frac{1}{r}\log(1 - w)\right)\right)\right]. \quad (4.7.10)$$

For embedding the characteristic function (4.7.10) in the class of characteristic functions corresponding to v-unimodal probability distributions we work as follows.

We suppose that the random variable X follows the exponential distribution with distribution function

$$F_X(x) = 1 - e^{-\mu x},$$

the random variable N follows the degenerate distribution with probability generating function $P_N(z) = z^n$ and we set $v = \mu n/r$, then (4.7.10) implies that the characteristic function of the stochastic model

$$Y = L\frac{1 - \mathrm{e}^{-rT}}{r}$$

has the form

$$\varphi_Y(u) = v\int_0^{1} P_S(\varphi_C(^{uw}/_{r}))(1 - w)^{v-1}dw \quad (4.7.11)$$

and the characteristic function (4.7.11) corresponds to a v-unimodal probability distribution.

An application of the stochastic model

$$Y = L\frac{1 - e^{-rT}}{r},$$

where $L = C_1 + C_2 + \cdots + C_S$ and $T = \min(X_1, X_2, \ldots, X_N)$in the area of considering an organization under conditions of competing risks is the following.

We suppose that the random variable N denotes the number of risks threatening an organization at the time point 0 and the random variable X_n denotes the time of the occurrence of the nth risk, the random variable $T = \min(X_1, X_2, , X_N)$ denotes the minimum risk occurrence time. We suppose that the random variable S denotes the number of different incomes that the organization creates at any time point of the interval $[0, T]$.

The random variable C_s denotes the size of the sth income that the organization creates at any time point of the interval $[0, T]$.

Hence the random variable $L = C_1 + C_2 + \cdots + C_S$ denotes the total size of different incomes that the organization creates at any time point of the above interval. The stochastic model

$$Y = L\frac{1 - e^{-rT}}{r}$$

denotes the present value, at the time point 0, of the total size of different incomes that the organization creates in the interval $[0, T]$.

Another application of the stochastic model

$$Y = L\frac{1 - e^{-rT}}{r},$$

where $L = C_1 + C_2 + \cdots + C_S$ and $T = \min(X_1, X_2, \ldots, X_N)$, in considering a firm under conditions of competing risks arises if we interpret the random sum $L = C_1 + C_2 + \cdots + C_S$ in the following way. We suppose that the random variable S denotes the number of different cash flows that firm saves at any time point of the interval $[0, T]$.

The random variable C_s denotes the size of the sth cash flow that the firm saves at any time point of the time interval $[0, T]$.

The random variable $L = C_1 + C_2 + \cdots + C_S$ denotes the total size of different cash flows that the firm saves at any time point of the time interval $[0, T]$.

Hence the stochastic model

$$Y = L\frac{1 - e^{-rT}}{r}$$

denotes the present value, at the time point 0, of the total size of different cash flows that the firm saves in the time interval $[0, T]$. □

Bibliography

Abate, J., & Whitt, W. (1996). An operational calculus for probability distributions via laplace transformations. *Advances in Applied Probability, 28*, 75–113.

Ackoff, R. (1971). Towards a system of systems concepts. *Management Science, 17*, 661–671.

Alamatsaz, M. (1993). On characterizations of exponential and gamma distributions. *Statistics and Probability Letters, 17*, 315–319.

Andersen, P., Borgan, Q., & Keiding, N. (1993). *Statistical models based on counting processes*. New York: Springer.

Artikis, C. T., & Artikis, P. T. (2005). Poisson random sums in modelling operations for treatment of ongoing risk occurrences. *SPOUDAI, 55*(2), 32–47.

Artikis, C. T., & Artikis, P. T. (2007). Incorporating a random number of independent competing risks in discounting a continuous uniform cash flow with rate of payment being a random sum. *Journal of Interdisciplinary Mathematics, 10*, 487–495.

Artikis, C. T., & Artikis, P. T. (2007). Properties and applications of a stochastic multiplicative model. *Journal of Interdisciplinary Mathematics,10*, 479–486.

Artikis, C. T., & Artikis, P. T. (2007). Risk management operations described by a stochastic discounting model incorporating a random sum of cash flows and a random maximum of recovery times. *Journal of Statistics & Management Systems, 10*, 439–450.

Artikis, C. T., & Artikis, P. T. (2008). Bernoulli selecting processes and integral part models in establishing properties and applications of a discrete distribution. *Journal of Interdisciplinary Mathematics, 11*, 443–450.

Artikis, C. T., & Artikis, P. T. (2008). Certain classes of discrete distributions in modeling risk control operations and establishing a transformation for probability generating functions. *Journal of Interdisciplinary Mathematics, 11*, 141–150.

Artikis, C. T., Artikis, P. T., & Moshakis, J. I. (2008). Stochastic compounding models for continuous uniform cash flows arising in risk management. *Journal of Statistics & Management Systems, 11*, 277–301.

Artikis, P. T., & Artikis, C. T. (2004). Stochastic models in fundamental risk management operations. *International Review of Economics and Business, 51*, 207–219.

Artikis, P. T., & Artikis, C. T. (2005). Properties and applications in risk management operations of a stochastic discounting model. *Journal of Statistics and Management Systems, 8*, 317–330.

Artikis, P. T., Artikis, C. T., & Fountas, C. E. (2006). Discrete renewal and self decomposable distributions in modelling information risk management operations. *Journal of Statistics & Management Systems, 9*, 73–85.

Artikis, P. T., & Artikis, C. T. (2008). Random sums of integral part models in computer systems operations. *Journal of Discrete Mathematical Sciences & Cryptography, 11*, 209–217.

Artikis, P. T., & Artikis, C. T. (2008). Thinning of renewal processes in stochastic discounting models and risk frequency reduction operations. *Journal of Interdisciplinary Mathematics, 11*, 291–300.

C. Artikis and P. Artikis, *Probability Distributions in Risk Management Operations*, Intelligent Systems Reference Library 83, DOI 10.1007/978-3-319-14256-2

Artikis, P. T., Artikis, C. T., & Moshakis, J. (2008). Discounted maximum of a random number of random cash flows in optimal decision making. *Journal of Information & Optimization Sciences, 29*, 1193–1201.

Artikis, P. T., Artikis, C. T., Agorastos, K. A., & Vlachos, A. (2009). Stochastic derivation and application in information risk frequency reduction of a class of discrete random variables. *Journal of Statistics & Management Systems, 12*, 59–64.

Baglini, B. (1983). *Global risk management*. New York: Risk Management Publishing Inc.

Baker, R. (1986). Handling uncertainty. *International Journal of Project Management, 4*(4), 205–210.

Barlow, D. (1993, April). The evolution of risk management. *Risk Management, 40*, 38–45.

Barlow, R., & Proschan, F. (1965). *Mathematical theory of reliability*. New York: Wiley.

Batabyal, A. (2001). Aspects of the theory of financial risk management for natural disasters. *Applied Mathematics Letters, 14*, 875–880.

Beck, U. (1991). *The risk society*. London: Sage.

Bernstein, P. (1996). *Against the gods: The remarkable story of risk*. New York: Wiley.

Berny, J. (1989). A new distribution for risk analysis. *Journal of the Operational Research Society, 40*, 1121–1127.

Bessis, J. (1998). *Risk management in banking*. Amazon.com. Inc.

Blum, J. R., & Rosenblatt, M. (1959). On the structure of infinitely divisible distributions. *Pacific Journal of Mathematics, 9*, 1–7.

Burges, D. (1983). Models and modelling. *Bulletin of the Institute of Mathematics and its Applications, 19*, 177–179.

Burlando, T. (1994, April). Chaos and risk management. *Risk Management, 41*, 54–61.

Carter, N., & Doherty, N. (1974). *Handbook of risk management*. London: Kluwer.

Checkland, P. (1972). Towards a system-based methodology for real-world problem solving. *Journal of Systems Engineering, 3*, 87–116.

Checkland, P. (1981). *Systems thinking, systems practice*. Wiley: Chichester.

Christoph, G., & Schreider, K. (1998). Discrete stable random variables. *Statistics and Probability Letters, 37*, 243–247.

Churchman, C. (1979). *The systems approach and its enemies*. New York: Basic Books.

Coleman, R. (1974). *Stochastic processes*. London: George Allen and Unwin Ltd.

Constandanche, G. (2000). Models of reality and reality of models. *Kybernetes, 29*, 1069–1077.

Cox, S., & Tait, N. (1991). *Reliability, safety and risk management—An integrated approach*. Oxford: Butterworth–Heinemann.

Crockford, N. (1980). *An introduction to risk management*. Cambridge: Woodward-Faulkner.

Cummings, T. (1980). *Systems theory for organization development*. Chichester: Wiley.

Damasio, A. (1994). *Descartes' error: Emotion, reason, and the human brain*. New York: Putnam Publishing.

Denney, D. (2005). *Risk and society*. London: Sage Publications.

Devroye, L. (1993). A triptych of discrete distributions related to the stable law. *Statistics and Probability Letters, 18*, 349–351.

Dharmadhikari, S., & Joag-Dev, K. (1988). *Unimodality, convexity and applications*. Boston: Academic Press.

Ekeland, I. (1993). *The broken dice and other mathematical tales of chance*. Chicago: University of Chicago Press

Evan, C. (1979). *The mighty micro*. London: Victor Gollancz Ltd,.

Eriksson, D. (2003). A Framework for the constitution of modelling processes: A proposition. *European Journal of Operational Research, 145*, 202–215.

Feller, W. (1966). *An introduction to probability theory and its applications* (2 ed., Vol. 1). New York: Wiley.

Fischoff, B., Watson, S., & Hope, C. (1984). Defining risk. *Policy Sciences, 17*, 123–129.

Flesch, R. (1966). *The new book of unusual quotations*. New York: Harper & Row, Publishers Inc.

Forrester, J. (1968). *Principles of systems*. Cambridge (Mass): MIT Press.

Forsam, E., & Sapatinas, T. (1995). Characterizations of some income distributions based on multiplicative damage models. *Australian Journal of Statistics, 37*, 89–93.
Forst, G. (1979). Characterization of self-decomposable probabilities on the half-line. *Zeitschrift fur Wahrscheinlichkeitstheorie und Verwandte Gebiete, 49*, 349–352.
Franklin, J. (1983). Philosophy and mathematical modelling. *Teaching Mathematics and Its Applications, 2*, 118–119.
Fromm, E. (1968). *The revolution of hope: toward a humanized technology*. New York: Harper & Row.
Fromm, E. (1973). *The anatomy of human destructiveness*. New York: Holt, Rinehart & Winston.
Gallagher, R. (1956). Risk management: New phase of cost control. *Harvard Business Review, 34*, 75–86.
Gerber, H. U. (1992). On the probability of ruin for infinitely divisible claim amount distributions. *Insurance Mathematics and Economics, 11*, 163–166.
Gibson, R. K. (1991, April). Making risk management happen in your organization. *Risk Management, 38*(4), 71.
Giddens, A. (1991). *Modernity and self-identity*. Cambridge: Polity Press.
Gollier, C. (2001). *The economics of risk and time*. Cambridge: MIT Press.
Gomes, M., & Pestana, D. (1978). The use of fractional calculus in probability. *Portugaliae Mathematica, 37*, 259–271.
Goovaerts, M., Vandebroeck, M., & Kaas, R. (1986). Ordering of risks and weighted compound distributions. *Statistica Neerlandica, 40*, 273–282.
Grandell, J. (1997). *Mixed poisson processes. Monographs on statistics and applied probability* (Vol. 77). London: Chapman & Hall.
Grose, V. (1978). *Managing risk*. Englewood Cliffs: Prentice Hall.
Haimes, Y. (1998). *Risk modelling, assessment and management*. New York: Wiley.
Head, G. (1984). *The risk management process*. New York: Risk and Insurance Management Society Inc.
Hertz, D. B., & Thomas, H. (1983). *Risk analysis and applications*. New York: Wiley.
Holloway, C. (1979). *Decision making under uncertainty: Models and choices*. Englewood Cliffs, NJ: Prentice-Hall.
Huizinga, J. (1955). *Homo ludens: A study of the play-element in culture*. Boston: Beacon Press.
Jackson, M. (1991). *Systems methodology for the management sciences*. New York: Plenum Press.
Jastrow, R. (1977). *Until the sun dies*. New York: Warner Books.
Jerwood, D. (1974). The cost of a carrier borne epidemic. *Journal of Applied Probability, 11*, 642–651.
Johnson, N., & Kotz, S. (1969). *Discrete distributions*. New York: Wiley.
Johnson, N., & Kotz, S. (1982). Developments in discrete distributions, 1969–1980. *International Statistical Review, 50*, 71–101.
Johnson, N., Kotz, S., & Kemp, A. (1992). *Univariate discrete distributions*. New York: Wiley.
Johnson, N., Kotz, S., & Balakrishnan, N. (1994). *Continuous univariate distributions* (2nd ed., Vol. 1). New York: Wiley.
Johnson, N., Kotz, S., & Balakrishnan, N. (1995). *Continuous univariate distributions* (2nd ed., Vol. 2). New York: Wiley.
Kamppinen, M., & Wilenius, M. (2001). Risk landscapes in the era of social transition. *Futures, 33*, 307–317.
Kaplan, S., & Garrick, B. (1981). On the quantitative definition of risk. *Risk Analysis,1*, 11–27.
Keilson, J., & Steutel, F. (1974). Mixtures of distributions, moment inequalities and measures of exponentiality and normality. *The Annals of Probability, 2*, 112–130.
Kellison, S. (1991). *The theory of interest*. Boston: Irwin.
Kendall, D. (1957). Some problems in the theory of dams. *Journal of the Royal Statistical Society, B19*, 207–212.
Kervern, G., et Rubise, P. (1991). L' Archipel du Danger, Editions Economica, Paris.
Kervern, G. (1994). *Latest advances in cindynics*. Paris: Economica.

Kervern, G. (1995). *Elements fondamentaux des Cindyniques*. Paris: Economica.
Kervern, G., & Boulenger, P. (2007). *Cindyniques: Concepts et mode d'emploi*. Paris: Economica.
Khintchine, A. (1938). *On Unimodal Distributions, Bulletin of the Institute of Mathematics*, Mec University. Kouybycheff de Tomsk, *2*(2) 1–7.
Kloman, F. (1976). *The Risk Management Revolution*. Fortune.
Kloman, F. (1992). Rethinking risk management. *The Geneva Papers on Risk and Insurance,17*, 299–313.
Kloman, F. (1995, April). Risk and response: Beyond 2000. *Risk Management, 42*, 65–72.
Kloman, F. (1996, September). Autopoiesis. *Risk Management Reports, 23*(9), 1–2.
Kolsrud, T. (1986). Some comments on thinned renewal processes. *Scandinavian Actuarial Journal 1986*, 236–241.
Kotlarski, I. (1962). On groups of independent random variables whose product follows the beta distribution. *Colloquium Mathematicum IX*, 325–332.
Kotz, S., & Johnson, N. (1991). A note on renewal distributions for discrete variables. *Statistics and Probability Letters, 12*, 229–231.
Krishnaji, N. (1970). A characteristic property of the yule distribution. *Sankhya, A, 32*, 343–346.
Krishnaji, N. (1970). Characterization of pareto distribution through a model of under-reported incomes. *Econometrica, 38*, 251–255.
Lagadec, P. (1990). *States of emergency: Technological failures and social destabilization*. London: Butterworth-Heinemann.
Laha, R. G., & Rohatgi, V. K. (1979). *Probability theory*. New York: Wiley.
Littlewood, B., & Strigini, L. (1992). The risk of software. *Scientific American, 267*(5), 38–43.
Lukacs, E., & Laha, R. (1964). *Applications of characteristic functions*. London: Charles Griffin.
Lukacs, E. (1969). A characterization of stable processes. *Journal of Applied Probability, 6*, 409–418.
Lukacs, E. (1970). *Characteristic functions*. London: Griffin.
Marshall, C. (1997, August). RiskWeb comment. *Risk Management Reports, 24*(8), 1–2.
McNeil, A. J., Frey, R., & Embrechts, P. (2010). *Quantitative risk management: Concepts, techniques, and tools*. Princeton, NJ: Princeton University Press.
Medgyessy, P. (1967). On a new class of unimodal infinitely divisible distribution functions and related topics. *Studia Scientiarum Mathematicarum Hungarica, 2*, 441–446.
Medgyessy, P. (1972). On the unimodality of discrete distributions. *Periodica Mathematica Hungarica, 2*, 245–257.
Meyer, W. (1984). *Concepts of mathematical modelling*. New York: Mc Graw-Hill.
Mitchell, T., & Harris, K. (2012, January). Resilience: A risk management approach. London: Background Note, Overseas Development Institute.
Mohan, N., Vasudeva, R., & Hebbar, H. (1993). On geometrically infinitely divisible laws and geometric domains of attraction. *Sankhya A, 55*, 171–179.
Morin, E. (1973). *Le Paradigme Perdu: La Nature Humaine*, Editions du Seuil.
Morin, E., & Kern, A. B. (1993). Terre Patrie, Editions du Seuil.
Murthy, D., Page, N., & Rodin, E. (1990). *Mathematical modelling*. Oxford: Pergamon Press.
Olshen, R., & Savage, L. (1970). A generalized unimodality. *Journal of Applied Probability, 7*, 21–34.
Pakes, A. (1992). A Characterization of gamma mixtures of stable laws motivated by limit theorems. *Statistica Neerlandica, 46*, 209–218.
Parzen, E. (1962). *Stochastic processes*. San Francisco: Holden-Day.
Passchier, W., & Reij, W. (1997). Risk is more than just a number, Vol. 8. RISK: Health, Safety & Environment.
Patil, G., & Joshi, S. (1968). *Dictionary and bibliography of discrete distributions*. Edinburgh: Oliver and Boyd.
Perry, J. (1986). Risk management—an approach for project managers. *International Journal of Project Management, 4*(4), 211–216.
Quantarelli, E. L. (1976). *Disasters: Theory and research*. California, Sage: Beverly Hills.

Ramachandran, B. (1994). Identically distributed stochastic integrals, stable processes and semi-stable processes. *Sankhya A, 56*, 25–43.
Ramachandran, B. (1997). On geometric stable laws, a related property of stable processes and stable densities of exponential one. *Annals of the Institute of Statistical Mathematics, 49*, 299–313.
Rao, B., & Janardan, K. (1985). An analog of the Rao–Rubin condition for distributions other than the poisson. *Pakistan Journal of Statistics, 1*, 1–15.
Rao, C., & Rubin, H. (1964). On a characterization of the poisson distribution. *Sankhya, Series A, 26*, 295–298.
Rao, C., Srivastava, R., Talwalker, S., & Edgar, G. (1980). characterization of probability distributions based on a generalized Rao–Rubin condition. *Sankhya, Series A, 42*, 161–169.
Rasmussen, J., & Green, A. (1982). *Human reliability in risk analysis in high risk safety technology*. New York: Wiley.
Rowe, W. (1988). *An anatomy of risk*. New York: Wiley.
Ross, S. (1970). *Applied probability models with optimization applications*. San Francisco: Holden-Day.
Sandquist, G. (1985). *Introduction to systems science*. Englewood Cliffs, NJ: Prentice-Hall.
Sapatinas, T. (1995). Characterizations of probability distributions based on discrete p-monotonicity. *Statistics and Probability Letters, 24*, 339–344.
Sapatinas, T. (1999). A characterization of the negative binomial distribution via a-monotonicity. *Statistics and Probability Letters, 45*, 42–53.
Settembrino, F. (1994, August). Risk management in enterprise: A systemic approach. *Risk Management, 41*, 34–37.
Shapiro, A., & Titman, S. (1985). An integrated approach to corporate risk management. *Midland Corporate Finance Journal, 3*, 41–56.
Skellam, J. (1958). On the derivation and applicability of Neyman's type a distribution. *Biometrica, 45*, 32–36.
Soong, T. (1981). *Probabilistic modelling and analysis in science and engineering*. New York: Wiley.
Steutel, F., & van Harn, K. (1979). Discrete analogues of self decomposability and stability. *Annals of Probability, 7*, 893–899.
Steutel, F. (1988). Note on discrete a-unimodality. *Statistica Neerlandica, 48*, 137–140.
Steutel, F., & van Harn, K. (2004). *Infinite divisibility of probability distributions on the real line*. New York, Basel: Marcel Dekker Inc.
Szekli, R. (1988). A note on preservation of self-decomposable under geometric compounding. *Statistics and Probability Letters, 6*, 231–236.
Szent-Györgyi, A. (1970). *The crazy ape*. New York: Philosophical Library.
Taylor, H., & Karlin, S. (1984). *An introduction to stochastic modelling*. Orlando: Academic Press.
Tijms, H. (1988). *Stochastic modelling and analysis: A computational approach*. New York: Wiley.
Tucker, H. G. (1967). *A graduate course in probability*. New York: Academic Press.
Wahlstrom, B. (1994). Modelling and modellers: An application to risk analysis. *European Journal of Operational Research, 75*, 477–487.
Wilde, J.S.G. (1994). *Target risk*. Toronto: PDE Publications.
Wilkinson, I. (2001). Anxiety in a risk society. Amazon.com.Inc.
Williams, C. A. Jr, & Heins, R. (1985). *Risk Management and Insurance*. New York: McGraw-Hill International Editions.
Yannaros, N. (1988). Some comments on the inverse problem for thinned renewal processes. *Scandinavian Actuarial Journal 2008*, 113–116.

Printed in the United States
By Bookmasters